Cable Communications Technology

Eugene R. Bartlett

McGraw-Hill

New York   Chicago   San Francisco   Lisbon
London   Madrid   Mexico City   Milan   New Delhi
San Juan   Seoul   Singapore   Sydney   Toronto

Copyright © 2005 by The McGraw-Hill Companies, Inc. All rights reserved. Printed in the United States of America. Except as permitted under the Copyright Act of 1976, no part of this publication may be reproduced or distributed in any form or by any means, or stored in a database or retrieval system, without the prior written permission of the publisher.

1 2 3 4 5 6 7 8 9 0   DOC/DOC   0 1 9 8 7 6 5

ISBN 0-07-145781-X

The sponsoring editor for this book was Jane Brownlow and the production supervisor was Azadeh Poursepanj. It was set in Century Schoolbook by G & S Book Services.

Printed and bound by R. R. Donnelley and Sons.

Throughout this book, trademarked names are used. Rather than put a trademark symbol after every occurrence of a trademarked name, we use names in an editorial fashion only, and to the benefit of the trademark owner, with no intention of infringement of the trademark. Where such designations appear in this book, they have been printed with initial caps.

CONTENTS

ABOUT THE AUTHOR

Eugene R. Bartlett has bachelor's and master's degrees from Northeastern University in Boston. Mr. Bartlett is a Professional Engineer and holds an FCC Radio Broadcast License. After graduation, Mr. Bartlett was on the research staff at Northeastern University and an assistant professor teaching engineering courses. In 1971 Mr. Bartlett moved with his family to Nantucket Island, 30 miles off Cape Cod, where he started his own business of marine electronics, radio, and television sales and service. Dismayed by the poor television reception on the island, Mr. Bartlett and three other island residents formed Nantucket Cablevision Corporation. After the system was built and activated, island residents enjoyed decent television. This 21-channel system was state-of-the-art in its day, with an active reverse system to pick up programming for rebroadcasts downstream to the subscribers. Seeing the need for other systems, Mr. Bartlett and other partners formed Bay Cable Affiliates; the company designed and built modern cable television systems that served many Massachusetts communities. When the last system was sold, Mr. Bartlett continued to perform consulting services. Following a move to Florida, Mr. Bartlett returned to teaching and taught communication courses at ITT Technical Institute. Mr. Bartlett accepted a faculty position at Valencia Community College in Orlando, Florida, as a Professor of Engineering and Technology, teaching courses in communication systems, and continuing as an Adjunct Professor of Engineering Technology.

ACKNOWLEDGMENTS

As I have in my past books, I thank my wife, Mona, for her helpfulness and dedicated work on the word processor. Also, again I thank my son-in-law, Dan Begeman, who handled the graphics even with his busy schedule as a commercial aircraft pilot.

I also wish to thank Mr. Bob Stark and in particular Mr. Ken Craig of Sunrise Telecom Broadband for providing information on cable communications instrumentation and testing in Chapter 7. The instruments Sunrise Telecom Broadband manufacture are state-of-the-art, battery-operated, and easy for technicians to use. They provide accurate results and have the ability to store information.

My colleagues at Valencia Community College, Dr. Gaby Hawat and Dean Nasser Hedayat, have been very supportive in producing this book. The faculty and staff of the Department of Architecture, Engineering, and Technology have also provided encouragement.

The Central Florida Chapter of the Society of Cable Telecommunications Engineers, where I have been a participating member for several years, has kept me up-to-date on the technical progress of our industry.

Eugene R. Bartlett, P.E.
Adjunct Professor
Valencia Community College
Orlando, Florida

PREFACE

This book is a textbook, designed to accompany a course in cable tele-communication technology. Students taking this course should have completed community college–level courses in algebra, trigonometry, and d.c. and a.c. circuit theory. Courses in signal filtering, amplifier theory, and basic transmission lines are also required to better grasp the fundamentals of cable communication technology. Cable communication technicians, many who have grown up in the business, are edging toward retirement, and many new technicians are needed. The Society of Cable Telecommunications Engineers advocates that technical schools and community colleges offer courses pertinent to the industry. Few if any schools today offer courses in cable telecommunications; perhaps for this reason, there are no appropriate textbooks. Several textbooks on the market devote a chapter or parts of a chapter to cable television system technology, but the treatments are incomplete. I hope this book will break the trend. In the meantime, cable communications networks are developing their technology and services.

Much of this book is taken from *Cable Television Handbook,* published in 2000. Chapter 2 has been divided into Chapters 2 and 3 in this book. Chapter 4 discusses the subject of fiber optics, and Chapter 5 covers digital technology for both television and telephone systems. Subscriber installation methods and subscriber terminal devices are handled in Chapter 6. Testing and maintenance requirements and techniques are contained in Chapter 7. Chapter 8, a new chapter, discusses the interconnection of telephone, cable, and LAN systems.

A technology some consider the "last mile," RF-type cable television signals carry both digital and analog services. The FCC has set rigorous standards for the control of signal leakage for this type of cable technology. To accomplish the required control, constant system testing is necessary. Many instrument manufacturers provide a variety of innovative test instruments that are easy to use and provide accurate measurements. Most companies adhere to an ongoing testing program; the tests are conducted by service technicians during ordinary service calls. In the near future, optical fiber may connect services to homes or business premises.

Telecommunication service providers view an all-optical system as the ultimate goal. This is not as close as some think because many new optical terminal devices have yet to be developed. One of the main goals of the cable telecommunications industry is to improve system reliability. Most of us know system reliability depends on many factors. AT&T's goal was to be in service 99.999% of the time, which amounts to approximately 5¼ minutes out of service per year. Clearly, this is extremely reliable. In order to achieve such great reliability, the telephone industry set up its network to carry its own electrical power. Cable television systems also carry their own power, and many systems employ standby power supplies in case of a commercial power interruption. This action has improved dependability immensely.

The cable telecommunications industry consists of the telephone company groups that have added high-speed digital signal services as well as plain old telephone service (POTS). High-speed DSL and ADSL technology have made it possible for these groups to offer video and audio as well as Internet service to their customers. Cable television operating companies can provide high-speed digital communications and telephone service with voice-over-the-Internet protocol (VOIP). With the addition of optical fiber technology, many project a conversion of services available to the public. Services and pricing will most likely determine which operator gets the business.

Introduction to Cable Communications Systems

Objectives

After learning the material in this chapter, the student will be able to

- Explain the forces that started cable television systems.
- Describe how television cable systems developed and evolved.
- Describe the improvement in equipment methods and construction techniques.
- Discuss the early regulatory forces imposed on early cable systems.
- Explain the concepts of cable loss and amplifier gains.
- Describe the units of measure and calculations to predict signal level performance.

1.1 Cable Television History

Cable television systems were first developed to improve television service to people living in remote areas where the off-air signal, or reception, was poor. Television broadcasting stations looked on this new industry as an ally in expanding their market area. Even today local broadcast stations are a major source of local news, weather, and sports to the viewing American public, and cable television still carries local television broadcast stations as part of its channel lineup.

Of course, the problem with reception for many television broadcast stations has not disappeared. Transmitted power output as stated in the station licenses has not changed. The location of the station transmitting antenna towers and antennas, in most cases, has not changed. The point is, the off-air reception problem remains, and therefore many subscribers still connect to the cable television system because it carries local television broadcasting stations. Many early cable television systems were required as part of their license, charter, or franchise to provide locally originated programs, providing another service to the subscriber for local news, weather, and sports. Many families were able to watch their children participate in school sports programs televised by the local cable company, shows available only to subscribers. Cable operators looked upon their locally originated cable channel as a horrible problem and large expense, though some became successful enough to sell local advertising spots to offset production costs.

Today, the cable television industry is facing much competition for the viewing public's attention. Analysis of the facts tells us that a family that has a direct satellite broadcast system (DBS) or some other satellite service has to have some kind of rooftop antenna or the usual rabbit-ear antenna to provide the local television off-air broadcasts. Many new housing developments and planned communities prohibit backyard dishes and/or rooftop antennas. These are additional reasons for the viewing public to remain on a cable television system. We, the cable operators, must never forget these facts.

1.11 The Television Reception Problem

The reception problem of off-air television broadcast stations remains; it has not simply disappeared with technical progress. Of course, the television set itself has been improved significantly. The development of solid-state components and the manufacture of printed circuits allowed sets to run cooler, thus prolonging their lives. Tuner circuits have improved signal-to-noise ratios, and the mechanical switch has been replaced by push-button digital switches. Remote controls have helped to make "couch potatoes" of many people. Large screens and improved sound have made modern television sets an important source of home entertainment.

1.111 Cable television is the major source of television programming in remote areas. Such areas often have a sparse population, so subscriber numbers are usually small. Cable systems serving such areas provide a very useful service and have to charge subscribers accordingly. Since the path loss of signal between the television transmitter and receiver is a function of distance and frequency, these areas receive weak television signals.

1.112 Terrain also affects signal levels. Mountainous areas cause problems with reception in many parts of the United States. It's no surprise, then, that the first cable television systems were started in this region. People living in the valleys, where most of the towns were concentrated, had major problems with television reception. Essentially, these locations were out of the field strength path as projected by the transmitting sites. In many instances, homes that were fairly close to the transmitter were nearly completely shielded by a mountain in between. Cable systems serving these areas enjoyed a high penetration of television service.

1.113 Urban residents also have television reception problems. Tall buildings and skyscrapers are similar to mountains in that they shield many apartment dwellers from good quality television reception. In many instances television signals are reflected off some of the surrounding buildings, causing severe television picture impairment called multipath reception. Multipath reception can be so severe that within the same home, the signal can be unwatchable for one channel and satisfactory for another. Such multipath problems are illustrated in Figure 1-1.

1.12 Early Cable Systems

Now that the need for cable television systems has been addressed, we will now briefly study the development of the industry. The technique used by early system operators was to start with good television off-air signals—usually received from a mountaintop, tower, or water tank location—and connect subscribers to this source by a cascade of amplifiers connected by sections of coaxial cable. Surplus coaxial cable from World War II and home-built radio frequency (RF) amplifiers made up the basic components of these early systems.

1.121 The preceding describes the system for what it actually was, a community antenna television (CATV) system, referred to in present-day vernacular as "classic cable." Such so-called start-up systems followed no

Figure 1-1
Multipath reception

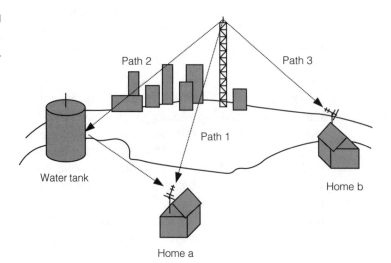

specified procedures in connecting the system. Technicians and engineers were usually fresh out of the armed services. Engineers who graduated from engineering schools were taught electronics as radio technology and engineering as power systems. Writings from telephone company sources such as AT&T/Bell Labs produced early theory on cascaded amplifiers. With such little information to go on, it was the resourcefulness and determination of the early system technicians and engineers that caused the industry to grow into what it is today. For many of us, our first systems were essentially classic cable systems that provided vastly improved broadcast television to our subscribers.

1.122   Most early classic systems simply consisted of a cascade of amplifiers connected together by sections of 75-ohm cable. Connectors were the kind used for that type of cable with the matching connectors on the amplifier housing. Matching networks were used to connect subscribers via a drop cable. The matching network tapped a signal off the main cable, feeding signal to the connecting drop cable. Amplifiers were usually either homebuilt or whatever was available on the newly formed master antenna television (MATV) industry. Power for these first amplifier types was often 110 volts alternating current (v.a.c.), thus requiring the power company to provide a power source. There was definitely a safety hazard to technical personnel working on these start-up systems.

The attenuation of the cable was not as good as the cable used today. Since the loss was much less at the low VHF bands, the upper VHF channels and any UHF channels were converted to the low VHF band of channel 2 through 6. This technique allowed more distance between amplifiers and thus provided more "reach" for the system. Many early systems were essentially single-cable, five-channel systems that enabled the start-up of the cable television industry.

1.123   The braided-type cable gave many system operators a lot of trouble, mostly with water ingress. Of course early systems leaked signal, but it was not a large problem since the signal leakage was "on channel." Noise ingress and the accompanying impedance mismatch, and picture ghosting were much more of a problem with system signal levels. Clearly, better cable was needed and since the supply of government-surplus cable became scarce, cable manufacturers eager for new business made improved cable for the fledgling industry. The solid-aluminum-sheathed, plastic-foam-filled cable fulfilled the cable television industry's needs. Several sizes of cable with improved loss characteristics allowed cable systems to extend the aerial plant longer distances, serving more subscribers. In addition to the cable, connector manufacturers were able to supply improved connectors, which exhibited improved electrical and weather-proofing characteristics. About the same time several manufacturers started making RF cable amplifiers with vastly improved noise figures, distortion characteristics, and improved operating-control features. These improvements caused the industry to grow at a rapid rate. The pressure tap was introduced with the arrival of the solid-aluminum-sheathed cable. This allowed a better method to connect subscribers to the system. These pressure taps exhibited an improved impedance match to the cable as well as better weatherproofing characteristics.

1.13 The Early Systems Evolve

Indeed, early systems were growing both in plant size and the accompanying subscriber population. In many instances, these systems made the difference between watching television and not watching television. Since more and more community residents wanted to get connected to the cable system, cable operators had to extend the plant. The improved cable, with its superior loss characteristics, enabled the use of the VHF high band, thus adding channels 7 through 13. The 12-channel system now gave cable operators more channels, thus causing an increase in subscriber interest. System architecture needed to be improved to cover a larger area and serve many more subscribers. Once the industry became an obvious success, more technicians and engineers became involved, both at the system and the manufacturing level. It is interesting to note that in many instances, as early literature indicates, the manufacturers were telling the cable operators how to extend and improve their systems and what was needed to do so.

The shortage of trained technical personnel became more and more of a problem. Realizing this, the Society of Cable Television Engineers was formed in the late 1960s and early 1970s. The society provided an exchange of technical information and practices among its members and eventually, through meetings and seminars, became the most important means of technical education. These technical meetings and seminars allowed the development of a great working relationship with many manufacturers and equipment vendors, who provided both speakers and technical literature to the members. The organization, now called the Society of Cable Telecommunication Engineers, continues to foster the education of its members. Its technical certification program provides a realistic measurement of its members' technical expertise.

1.131　The evolution of network architecture was one of the main developments of early cable television systems. From the simple single-cable system, the trunk-feeder system was developed. Cascade amplifier theory indicated that if signal level was preserved by limiting the losses attributed to subscriber taps, the overall distance or reach of a system would be improved. At amplifier locations, a tap-off or bridging amplifier was added; a small amount of trunk signal was tapped off, amplified, and connected to a separate feeder cable system containing an improved subscriber tap. These taps were actually directional couplers with more than one output port. The main trunk cable carried a high-quality signal to the extremities of the cable system, while the feeder cable carried the subscriber cable with the taps. Subscribers were connected to the system via a drop cable connected to one of the tap ports. This concept allowed the cable operating companies to build their plant farther from the signal source or headend, passing more and more homes along the way. Higher and higher subscriber counts resulted in improved cash flow needed to support company operations and pay down debt.

1.132　This network topology resulted in what is called the trunk-feeder system or, more appropriately, the tree network. No subscribers were connected to the trunk cable, thus preserving the signal quality. Subscribers were only connected to the taps in the feeder network. Thus the "trunk and branch" likeness to the tree resulted. The design process of such a system is now a lot different from the early single-cable systems. This type of cable system is shown in Figure 1-2. Essentially, the trunk cable was a signal transportation system, and the feeder network was the signal delivery system.

1.133　Early system amplifiers were powered by the 110-v.a.c. line. Often a watt-hour meter was installed at one amplifier, and its power consumed was monitored by the local power company. Often the other amplifiers were merely counted, and the single amplifier consumption multiplied by the number of amplifiers was used to calculate the monthly power charges to the cable company. In some instances, the local power company required a meter at each amplifier location. Connections to the local power distribution system required the cable operators to properly ground and bond their systems according to the National Electric Safety codes. Of course, there was the possibility of electric shock.

1.134　Another improvement that had an immense impact on the cable television industry was the development of the directional tap. Also

Figure 1-2

Example of a trunk-feeder design

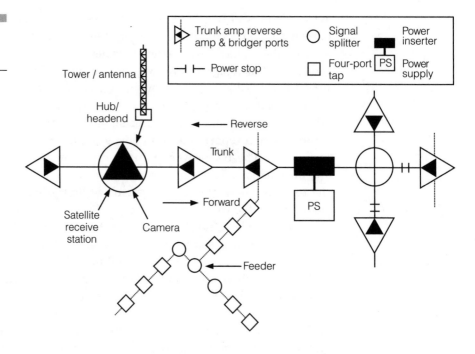

known as the subscriber tap, it provides signal to the drop cable feeding the subscriber's television set. These taps were made with an input and output coaxial connector for the feeder cable, plus two or more "F" connector ports for the subscriber connections. The "F"-type connector and the "C"-type connectors were spin-offs from the MATV industry. Today many people in the cable television industry wish the 75-ohm version of the BNC connector was adopted instead of the "F"-type connector. However, present-day systems use a vastly improved "F" connector that, when properly installed, works very well.

1.2 Changes in Plant Design and Development

Changes in plant design and network architecture were driven by the development of the devices and components from the various manufacturers and equipment vendors. This equipment often pointed the way to the possibilities of such network improvements. Across the board, cable, amplifiers, couplers, dividers, taps, connectors, and power supplies were vastly improved—and they still are even today. Each of these devices will be examined in the remainder of this chapter.

1.21 Electronic Equipment Development

The development in electronics always seems to be a source of amazement. Early amplifying devices were the vacuum tube that consumed much more power than the present-day solid-state devices. The development of the transistor and its use as an amplifying device was welcomed in the cable amplifier arena. The transistorized amplifiers consumed less power, developed less heat, and provided lower noise and distortion figures. It was a definite improvement. Solid-state technology progressed rapidly from the single discrete transistor to the integrated circuit consisting of the equivalent of several transistors as well as resistors and capacitors. Such so-called chip technology resulted in integrated circuit "IC" amplifiers for the cable television industry. These amplifiers provided more gain and a higher output level, as well as lower noise and distortion characteristics.

1.211   One large improvement in cable television technology was the use of cable-powered amplifiers. The cable itself was used to feed power to the electronic amplifiers. Coupling networks within the amplifier housings would separate the signal from the power, thus feeding each to their respective circuits. At first the voltage level of 30 v.a.c. was connected to the cable via a transformer with its primary joined to the 110-v.a.c. commercial power system. This 30-volt power would supply a segment of the cable system consisting of several amplifiers. Power blocks would disconnect the segment from other segments connected to other 30-volt supplies. These power blocks were bridged with signal-passing networks, thus providing signal continuity between power segments. Now, fewer connections had to be made to the commercial power system. Only at 30-volt power supply locations were connections to the commercial power system needed.

This improvement now requires system designers to consider the power-distribution problem along with the signal-delivery system. Systems were always expanding and new areas added, requiring more amplifiers and power supplies. It quickly became evident that increasing the cable power supply voltage to 60 volts would result in less line voltage drop and would allow more amplifiers per power segment. This voltage value falls within the National Electric Safety code as low voltage, and thus was often considered as the optimum voltage value. However, today some systems use supplies with 90-v.a.c. as the cable supply voltage of choice. In many of today's modern cable systems that employ fiber-optic technology, the fiber-optic system is powered from the coaxial cable section of the outside plant.

1.212 The solid-state cable system amplifiers evolved from the transistorized version to the integrated circuit type. The main amplifier chip as manufactured by several companies is complete enough to be considered an amplifier block. Companies using such chip technology merely have to supply the input/output and power connections plus several control connections for amplifier gain and slope. In some amplifier integrated circuits, temperature compensation and circuit stabilization have to be provided. Proper heat sinking for these integrated circuits became part of the mounting method within the amplifier housing. This technology contributed to improved system performance as well as lower cost.

In later years, the bandwidth of such amplifiers provided proper gain and output level up to and including 1000 MHz (1 GHz). This bandwidth of about 1 GHz allowed the operation of up to approximately 150, 6-MHz NTSC television channels. It became quite clear that this large downstream bandwidth was indeed overkill, and other uses may, in the end, become more profitable. Thus, the cable telecommunication concept can make good use of the overall bandwidth.

The amplifier downstream delivery of television signals provided service basically "on channel," or on the frequency of television broadcast stations. No additional equipment is needed by the subscriber. When programming was carried on the cable on nonstandard television channels, a means of converting the programming to channels that a television set could receive was needed. The amplifiers had no problem with nonstandard channels but the television sets did. That, of course, is another story.

1.213 Because lightning and power surges caused power supply and amplifier failure, protective devices were developed. In older systems, fuses were the first line of defense, followed by gas surge suppressors. Most systems in operation today use a variety of protective devices, but proper system design can make a power supply more immune to transients and electrical surges.

First and foremost, the ferroresonant transformer with a secondary winding connected to a large capacitive load provides a reasonably stable 60-v.a.c. supply. Usually the power inserter has protective fuses on the supply ports. At each amplifier location, a surge protector such as the gas variety or a metal oxide varistor (MOV) protected the amplifier power supply module from the cable system. Most modern cable amplifier power supplies were switching power supplies in which the voltage to the amplifier gain blocks (amplifier-integrated circuit) was highly regulated.

The previously mentioned ferroresonant transformer power supply was first developed as a 30-v.a.c. power supply. As many systems grew larger it became obvious that a higher voltage power supply was more applicable to the cable system powering needs. Hence, the 60-volt power supply became standard and remained so for many years. Now, with many cable plants using fiber-optic systems powered by the coaxial cable system and coupled with an activated upstream system, a 90-volt supply obviously was needed.

Since plant reliability is extremely important in cable systems designed for two-way voice/video and data service, standby power supplies providing an appropriate amount of standby time are definitely required. Because the nature of ferroresonant transformer supplies requires the core to remain in saturation, the output becomes a quasi-square wave. The primary current has to keep the transformer core in saturation at light loads, which means the current drawn is significant. Thus, standby supplies use a different, less power-hungry method of regulation needed to provide a decent amount of standby time. It is, of course, desirable to have the a.c. power supply operating between 75 and 85 percent full load at least.

1.214 Based on experience with early systems, the need for automatic gain control was quickly recognized. Variations in amplifier output due to temperature changes occurred in both the cable and the amplifiers themselves. Thermal compensation of the amplifier gain helped greatly in the stabilization of the amplifier output levels, but the compensation of signal levels due to cable temperature changes required automatic gain-controlled circuits. These circuits required two pilot frequencies located in both the low portion and the high portion of the signal band. The pilot carriers were appropriately termed the "high pilot" and the "low pilot." Level circuitry measuring the amplitude of both carriers responded by controlling the gain of the amplifiers at both the high and low end of the frequency band.

At one time it was thought that alternating between controlled amplifiers and non-gain-controlled amplifiers (manual gain controlled) was sufficient to maintain a constant output level. However, as greater bandwidths and higher levels were needed as systems expanded in distance and in the number of channels, every amplifier in a trunk cascade had automatic gain control (AGC), which then became the standard. As distance requirements for cable plants increased, better amplifiers were needed. Amplifiers that had high-level output, lower noise, and distortion specification were also developed. The feed-forward amplifier improved third-order distortion specification and was useful in long amplifier cascades where third-order distortion was a limiting factor. The power-doubling amplifier provided higher signal levels in long cascades and in feeder branches that contained a large amount of subscriber taps.

Many amplifier manufacturers provide a broad selection of amplifiers that aid system designers in designing efficient and economical cable plants. The power-doubling and feed-forward designed amplifiers enabled coaxial cable trunk lines to be extended farther into the system. Cable system operators could install the reverse amplifier modules into sections of the coaxial trunk to activate the sections desired for two-way service. The required extra system power needed to activate a trunk feed or return feed should be designed into the system in the first place to avoid having to upgrade the power system.

1.215 The cable system amplifiers in present-day systems are greatly improved over the previous type. Earlier amplifiers lacked proper thermal stability, causing the gain to shift. Thermal compensation devices that appeared in the newer designs took care of this problem. Since cable characteristics also change with temperature, AGC became a necessary addition to the amplifiers.

The AGC method employed the use of a low-frequency and high-frequency pilot whose signal level was used as a reference level for the automatic control of system signal level. Coupled with the built-in thermal compensation, the overall temperature problem was under control. Now, system signal levels for long trunk amplifier cascades were stabilized and subscriber signal levels did not vary.

As solid-state technology advanced into the integrated circuit area, amplifiers in cable television systems improved immensely. Less power was consumed, and thermal stability improved because much of the circuitry was contained in the same IC chip. Now with automatic gain and slope controls added to an improved amplifier, many cable systems could expand their bandwidth, thus increasing the number of channels offered to subscribers, which had a positive impact on cash flow. On the other hand, if it was more desirable to extend plant distance rather than expand bandwidth, the lower noise low distortion amplifiers made this trade-off possible. Then with the arrival of optical fiber, the whole picture changed again—for the better. This subject is covered in detail in Chapter 4.

1.216 The unit of cable system gain or loss is the dBmV. Amplifier and/or filter response is measured in logarithmic units. Commercial systems usually are measured in dBm where 0 dBm corresponds to a power level of 1 milliwatt (mW). In cable systems employing various devices the measurement unit of choice is the dBmV where 0 dBmV corresponds to a voltage level of 1 millivolt (mV). For the dBm the impedance is 50 ohms that consumes 1 mW. In the case of a cable system the impedance is 75 ohms that has 1 mV across it. At this time students should refresh their memory on the use of logarithms. Table 1-1 relates the power of 10 to logarithmic equivalents. A mathematical relationship exists between the voltage across a resistance, the current through the resistance, and the power consumed by the resistance.

$$\text{Power } (P) = \frac{V^2}{R}$$

Because $P = I^2R$ and by Ohm's law,

$$I = \frac{V}{R}$$

so

$$I^2 = \frac{V^2}{R^2} \quad \text{and} \quad P = \frac{V^2}{R^2} \times R = \frac{V^2}{R}$$

because dB is actually a power ratio in logarithmic notation. Mathematically stated,

$$dB = 10 \log \frac{P_{\text{output}}}{P_{\text{input}}} \tag{1-1}$$

For $P_{\text{out}} = V_{\text{out}}^2/R$ and $P_{\text{in}} = V_{\text{in}}^2/R$,

$$dB = 10 \log \frac{V_{\text{out}}^2/R}{V_{\text{in}}^2/R} = 10 \log \frac{V_{\text{out}}^2}{V_{\text{in}}^2}$$

Table 1-1

Logarithms, Power Ratios, and Voltage Ratios

Number	As Power of 10	Log Number	No. As a Power Ratio (in dB)	No. As a Voltage Ratio dB
10,000	10^4	4	40	80
1,000	10^3	3	30	60
100	10^2	2	20	40
10	10^1	1	10	20
1	10^0	0	0	0
0.1	10^{-1}	-1	-10	-20
0.01	10^{-2}	-2	-20	-40
0.001	10^{-3}	-3	-30	-30
0.0001	10^{-4}	-4	-40	-40

Note: When the number is expressed as a power ratio, $dB = 10 \log \frac{P_1}{P_2}$

When the number is expressed as a voltage ratio, $dB = 20 \log \frac{V_1}{V_2}$

A logarithmic identity states $\log n^2 = 2 \log n$. We may write

$$dB = 10(2)\log \frac{V_{out}}{V_{in}} \tag{1-2}$$

$$dB = 20 \log \frac{V_{out}}{V_{in}} \quad \text{since } R_{in} = R_{out} = R$$

Example 1-1

For $P_{in} = P_{out} = P$, using Equation 1-1,

$$dB = 10 \log \frac{P}{P} = 10 \log 1 = 0 \text{ dB—no loss, no gain}$$

For $P_{in} = 100P_{out}$,

$$dB = 10 \log \frac{P_{out}}{100P_{out}} = 10 \log \frac{1}{100} = 10 \log 0.01 = -20 \text{ dB}$$

This is a loss of 100:1, or -20 dB, where $(-)$ means a negative gain or a loss.

Example 1-2 For a voltage ratio using Equation 1-2,

$$V_{out} = V_{in} = V \text{ dB} = 20 \log \frac{V}{V} = 20 \log 1 = 0 \text{ dB}$$

$$V_{in} = 100V_{out}, \text{ dB} = 20 \log \frac{V_{out}}{100V_{out}} = 20 \log \frac{1}{100}$$

$$\text{dB} = 20 \log(0.01) = -40 \text{ dB}$$

This is a loss caused by a 100:1 voltage reduction. In cable systems a constant impedance matched system is maintained to 75 ohms. Therefore 0 dBmV is defined as 1 mV across 75 ohms. Using Equation 1-2, we may write

$$\text{dBmV} = 20 \log \frac{V}{1 \text{ mV}}$$

Example 1-3 If $V = 1$ mV,

$$\text{dBmV} = 20 \log \frac{1 \text{ mV}}{1 \text{ mV}} = 20 \log 1 = 0 \text{ dBmV}$$

Example 1-4 If $V = 10$ mV,

$$\text{dBmV} = 20 \log \frac{10 \text{ mV}}{1 \text{ mV}} = 20 \log 10 = 20(1) = 20 \text{ dBmV}$$

Table 1-2 illustrates the dBmV level versus the voltage ratio in a 75-ohm system. Now the student can use this information to calculate the gains (+dBmV) and losses (−dBmV) in a cable system.

1.217 It has been proven long ago and in practice that the unity gain building block is the concept that is used in cascaded amplifier technology. Depending on the amplifier's output specifications, the concept is to allow the cable loss to equal the amplifier output level and vice versa. Essentially, the amplifier cable section calculates to unity gains; in other words, the cable loss equals the amplifier's output level. The cable loss is the highest at the upper frequencies, so unity gain is calculated at the upper frequency limit. This concept is illustrated in Figure 1-3.

The low-frequency end of the spectrum will arrive at the amplifier input at a much higher signal level than the high-frequency end because the low frequencies are attenuated a lot less by the cable section. The sig-

	Voltage (in mV)	dBmV
1 V =	1000	60
	100	40
	10	20
	1	0
	0.1	−20
	0.01	−40
	0.001	−60

Note: dBmV = 20 log V (mV)/1 (mV)

1 mV across a resistance of 75 ohms = 0 dBmV.

Figure 1-3
The unity gain
building block

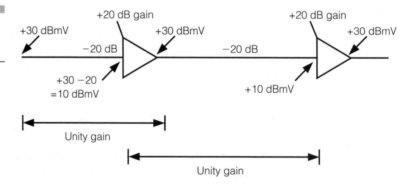

nal level at the low-frequency end of the spectrum has to be corrected by the internal equalizer network at the amplifier input section. Such equalizers come in a variety of values and are plug-in, so the correct value can be selected for the specific cable section. Present-day amplifiers still have selectable pads and equalizers to adjust the input signal level for each amplifier station.

The standard procedure is to measure the amplifier input signal level at both the low and high end (usually pilot frequencies) and then select the appropriate pad and equalizer to facilitate the setting of the output gain and slope. This technique will be treated in more detail later in the book. This procedure is known as amplifier or system balancing and is extremely important for the amplifier cascade to perform with optimum distortion and noise specifications. For a study of amplifier cascade noise see Appendix E.

1.22 Development of Coaxial Cables

Coaxial cable has undergone many developments over the years. The first type of coaxial cable had a braided copper wire shield and was covered by a plastic (polyethylene) jacket. Shielding effectiveness was poor because of the weaving of the braided shield, which allowed the cable to be quite flexible. Connecting to such a shield was a problem in itself. The center conductor was often a copper wire or copper-plated steel wire. The insulation was often a harder polyethylene. Early coaxial cable was usually found on the surplus market left over from World War II. This type of cable had high attenuation, poor shielding, and was easily damaged by moisture. With the start of the cable television industry, cable manufacturers were quick to realize the need for more and better cable.

1.221 The solid-aluminum-sheath cable was introduced in the 0.412-in and 0.500-in sizes with a nominal impedance of 75 ± 2 ohms. The dielectric was either Styrofoam or polyethylene foam, and the center conductor was either solid copper or copper-clad aluminum. A polyvinyl chloride (PVC) jacket was available and was recommended in coastal climates.

The solid-aluminum sheath (tubing) vastly improved the shielding effectiveness of the cable. In a great many areas, the bare aluminum cable worked well and became widely used. A whole series of connectors appeared on the market. Some were better than others and eventually the best became even better as the cable television industry grew. The solid aluminum sheath cable was difficult to handle; it was a lot stiffer and could be easily kinked, thus destroying the transmission capability. Proper construction procedures, tools, and equipment were developed to correctly handle and install the cable in an expeditious manner. With the development of this cable, the cable television industry could extend its plants, further serving more subscribers with more channels and improved pictures.

1.222 Various techniques were used to try to improve the loss-versus-frequency characteristics of coaxial cable. Since dry air has a velocity constant of 0.95, a cable with pure dry air as a dielectric should provide about the best attenuation-versus-frequency characteristic. Because some means has to be employed to keep the center conductor in the center (air itself just does not do the job), one cable manufacturer used polystyrene disks placed a few inches apart, sealing in cells of air as well as supporting the center conductor in a concentric condition. Such cable did provide superior loss-versus-frequency characteristics, but it was difficult to manufacture because fusing the disks in place was a critical process. Some of this cable is most likely still in service and an improved version is available today.

Coaxial cable used today uses the polyfoam technique, where the plastic foam forms bubbles of air that control the dielectric constant. This cable has proven to be very rugged and maintains its characteristics over most climatic conditions.

1.223 The attenuation of cable as a function of frequency is a well-known characteristic to the cable television industry. Essentially, the loss, or attenuation, versus frequency is a logarithmic function, as shown in Figure 1-4. Most manufacturers publish a chart of attenuation versus frequency so system designers can calculate the cable span loss between amplifiers. The usual design technique is to calculate the span loss at the highest operating frequency that exhibits the greatest attenuation. This assures that each amplifier has the proper input and output signal levels at the upper frequency limit. The lower-frequency end will have to use some equalization so the amplifiers are not overdriven. As far as cable performance, there have not been any large breakthroughs in cable loss characteristics, but several manufacturers have improved their manufacturing techniques, which have produced cable that is more rugged, has a lower return loss characteristic, and demonstrates improved environmental performance.

Essentially, the loop resistance of coaxial cable depends on the conductivity of the center and outer conductors. Since the solid aluminum sheath cable has high conductivity for the shield, most of the loop resistance is attributed to the center conductor. Making the center conductor out of aluminum and cladding it with copper produces a lower resistance at the RF operating frequencies due to the well-known skin effect. However, the resistance at 60 Hz is appreciable enough to limit cable powering of the repeater amplifiers. By using a solid-copper center conductor, the loop resistance of the cable span is decreased, resulting in possibly fewer 60-Hz power supplies.

This fact has to be taken into account when working out the cost per mile figures. Also, the larger the cable size, the larger diameter center conductor, which results in lower loop resistance values. Again, lower loop resistance caused by either choice, solid-copper center conductor or larger

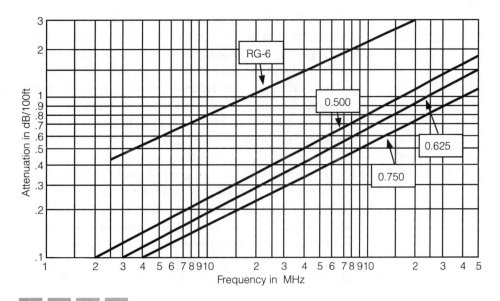

Figure 1-4
Cable attenuation versus frequency

cable size, can result in a fewer number of power supplies. Solid-copper center conductor cable and larger size cables are more expensive, so the cable designer has to weigh the pros and cons to arrive at the most cost-effective solution. It should be kept in mind that larger cable sizes can also result in greater spacing between amplifiers due to lower attenuation and hence could result in fewer amplifiers in a cable run.

1.224 For many applications in areas where there is no saltwater environment, bare aluminum-sheath cable wears well. Coastal areas or industrial areas where air pollution couples with a wet climate, however, could cause a corrosive problem with aluminum. For such general corrosive environments, jacketed cable should be used.

In general, jacketed cable is more rugged simply because of the added strength of the jacket. Jacketed cable is heavier than unjacketed cable and for the same size cable is larger in diameter. This can limit the amount of cable length that can be placed on a reel. However, this usually is not much of a problem because in practice the ends of a reel often are too short for use and simply discarded. This is often referred to as shrinkage of cable stock.

Coaxial cable that is manufactured for underground installations is always jacketed and contains a flooding compound between the plastic (polyethylene) jacket and the aluminum sheath. If a break, cut, or gash occurs in the plastic, the flooding compound flows into the damaged area and seals it against moisture ingress. The use of polyethylene plastic in jacketing the cable, as well as the plastic bubbled foam used as the dielectric, has been responsible for some of the major improvements in the cable television industry.

1.225   Through the years coaxial cable connectors have evolved from rudimentary to sophisticated, complex designs. One of the difficulties discovered early on was that many connectors had the problem of keeping a good electrical contact around the aluminum sheath. Loss of a complete connection around the circumference of the aluminum sheath caused RF energy to escape from the cable signal (called leakage) as well as electrical noise to enter the cable (called ingress), which in turn interfered with the cable signal.

Various techniques were used to assure good electrical connections for the center conductor and the surrounding aluminum sheath. Two types of connectors were developed. One was called the straight-through type, where the center conductor was carried through the connector to be fastened directly to the device being connected. The second type was referred to as the stinger type, where the center conductor entered a viselike grip socket. Once the center conductor entered the connector it could not be pulled out. This type of connector had a protruding pin called a stinger, which was connected to the center conductor. This pin was often trimmed to fit the connecting device.

The cable preparation for each of these connector types was different and had to be followed precisely to assure a proper fit. Several tool manufacturers made devices that would remove a portion of the jacket, core the cable dielectric, and cleave and clean the center conductors. Often the ends of the cable were prepared for a certain manufacturer's connector.

To obtain proper shielding characteristics, various methods were used. The cable was usually in some form of longitudinal stress due to its own weight, as well as wind and vibration. Such tension can cause the cable to pull out of the connector, thus breaking the connection and interrupting service. The lashing operation alleviated most of the tension, and the connector's grip on the cable sheath and center conductor prevented this from happening. To improve on the connector's grip on the aluminum cable sheath, a stainless steel sleeve was used to add strength to this grip. This sleeve fit inside the aluminum sheath while the connector's wedge-type grip forced the aluminum against the sleeve, improving the connector's grip on the aluminum sheath.

Initially, some stainless steel sleeves made for various cable sizes had to be installed first. The procedure was to remove some of the dielectric core, called coring, using a tool made for the cable size and specific connector and then inserting the sleeve. At present, most connector types have an integral sleeve placed within the connector that prevents losing a sleeve, installing an improper sleeve size, or forgetting to install a sleeve. An example of a connector type is shown in Figure 1-5 and the preparation steps are displayed in Figure 1-6. Proper connector installation prevents

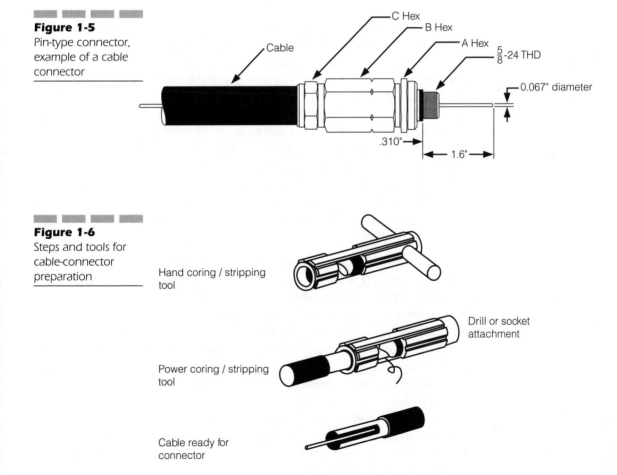

Figure 1-5
Pin-type connector, example of a cable connector

Figure 1-6
Steps and tools for cable-connector preparation

Hand coring / stripping tool

Power coring / stripping tool

Cable ready for connector

Drill or socket attachment

signal leakage, noise, and interference ingress as well as ensures good weatherproofing.

Since coaxial cable carries its own power, usually a 60-volt or 90-volt 60-Hz a.c., the cable connectors have to provide good current-carrying capacity referred to as "ampacity." Cable manufacturers cover the current carrying capacity of their cables along with other specifications in their catalogs and data books. The main thing is to select a cable type and an appropriate connector manufacturer and then to order the accompanying tools to prepare the cable and install the connectors.

1.23 Construction

Few cable system operators install their own cable systems. Most systems are constructed by cable construction contractors who provide equipment and crews to do the job. These contractors often use subcontractors as specialists, such as splicing (connectorizing) contractors. The cable operator has to require that, in the case of a prime or principal contractor, approval be given to all subcontractors before work begins. A cable operator should also require that co-insurance clauses and hold-harmless clauses be part of the contract with the prime contractor.

Proper construction practices, work progress, and contract performance milestones are important to the contract's completion timetable. The lending institution (bank) may require contract supervision be performed by an independent consultant. Contract monitoring is extremely important to the cable operator during the construction phase of the project.

1.231 An aerial plant is selected by a cable operator because the power and telephone services are provided by existing aerial pole lines. Rural areas are often served by utility pole-lines. Cable systems have to apply to the pole owners for permission to install the cable lines on the utility poles. These pole systems are either solely owned by the power utility or the telephone company or in many instances the poles are owned on a 50-50 basis, where the electric company and the telephone company each own half of the poles. Such pole plant is known as "joint own" or simply "J-O" poles.

Cable operators are leased space in a location usually 1 foot above telephone lines and 41 inches from the standard 120/220-volt power distribution service lines. The top section of each pole is reserved for high-voltage electric distribution service.

If there is not any space for a cable system, as determined by a three-party survey, then the cable operator is required to pay the cost for either a new or taller pole. If proper space is available but the electric and telephone systems have crowded the pole, then the cable operator is required to pay the respective companies to rearrange their plant to make space available for the cable system. The often difficult and time-consuming process of applying for use, surveying each pole applied for use, and re-arranging the plant is known in the cable industry as make-ready. No cable construction is permitted until this phase is completed and paid for. Once construction is permitted, work is allowed to proceed according to standard pole-line construction procedures, as written in the Telco Blue Book. Copies of this comprehensive construction manual are often made available to the cable operators by the telephone company to ensure that the plant is properly built. For overhead or aerial plants, this is the way it has been done and is being done today.

Present-day aerial pole-line plants are getting very crowded, and larger poles have been required for upgraded and increased telephone and electric services. In many cases, the fire alarm–wired systems have been replaced by wireless systems that require a lot less maintenance but retain high reliability. Removal of this plant has allowed more space for cable operators' expanded plants. The telephone companies, in many instances, are making additions to their plants using underground cable installation—and so are many cable operators.

Aerial plant equipment has been improved throughout the years. Line amplifier housings have improved metals with upgraded RFI gaskets, as well as weatherproof gaskets. More attention is being paid to the characteristics of the many types of metals being used together in aerial pole-line plants. Telephone companies have learned a lot from past experiences with dissimilar metals, grounding, and bonding. Proper grounding procedures, directed by the Telco Blue Book, have to be meticulously followed by cable construction contractors. The cable operator technical personnel should closely monitor the quality of construction of their new cable plant because they will have to live with it for years to come.

Progress in construction equipment has been forthcoming both in tooling and equipment. One of the most significant is the hydraulic lift truck. Prices have become more reasonable because more truck makers have been introducing their products to market. More used equipment is becoming available as well, so even smaller cable operators can afford to have at least one truck. Having a hydraulic lift truck available to a cable operator is important because a possibly dangerous situation can be handled safely using a bucket truck.

1.232 As mentioned earlier, placing a communications cable system underground is becoming more and more attractive. In heavily populated urban areas, underground facilities are often required. Where underground ducts with connecting vaults are available, cable facilities could be placed. The cable system is often required to lease duct space from the telephone company, electric company, or municipality. This often used to be the only choice, because no utility poles were available. Cable companies operating under these conditions had to have the necessary equipment, such as work area signs, barricades, vault ventilating blowers, and the like, in order to perform underground plant maintenance properly.

In suburban and more rural areas, direct burial of coaxial cable is a choice when soil conditions allow for it. Sandy soil, soft clay, or gravel free of stones is best for direct burial of cable. Trenching or plowing of the cable provided a quick and efficient method of directly placing cable underground with a minimum of restoration. Where the soil contains a lot of rocks and stones, placing the cable in plastic duct is the best method. At first, trenching was used and the cable was placed in plastic duct, laid in the trench, and then the joints were cemented. Splices, electronic equipment, taps, and power supplies were placed in either metal vertical pedestals or horizontal plastic or metal vaults.

Today, integral cable placed in flexible conduit is presently available for use in a variety of soil conditions and can be either plowed directly or placed in trenches. If several cables are to be installed at the same time, cable in conduit can be ordered in several color choices for easy identification later during repair situations.

1.233 As the cable industry matured and improved, the equipment did as well. Originally, pole-mounted cable construction was performed by technicians climbing poles with belts and hooks. Many of the cable construction people got their training from their former employer, such as the electric company, the telephone company, or the Army Signal Corps. Getting up and down a pole in a safe and expeditious manner was the name of the game. For several years, poles were climbed with belts and hooks; ladders tied at the poles were used by cable construction personnel. Then along came the lift vehicles. Some were simply electrically operated truck-mounted extension ladders and some were hydraulically operated jointed boom types. Both had cost and operational features.

The vehicles used by the electric companies were usually insulated for certain required high voltage levels and used insulated hydraulic control lines. Since cable operators were working above telephone lines and below power lines, they often used insulated trucks. The jointed hydraulic aerial basket truck was often the best choice because it provided a large amount of aerial positional articulation as well as the protection against shock hazards in case of accidental contact with electric lines. The equipment manufacturers were quick to realize the cable system applications and provided the industry with smaller and less expensive articulate, insulated lift vehicles. Cable operators were introduced to this equipment through local dealers and local trade shows. In short order, such equipment became available on the used market, making them available to cable operators with smaller budgets.

Some manufacturers of hydraulic lift trucks offered an accompanying variety of hydraulically operated tools. Drilling holes in poles, tightening the pole-line hardware, and raising equipment all became quicker and easier with the use of such tools. Cable construction companies using such tooling could produce quality work faster than those using manual methods.

Some subscriber drop installers use ladders instead of the belt-and-climber methods. These ladders should be made of fiberglass (insulated), placed at the pole, and strapped to the pole to prevent accidental slipping. For personnel untrained in the use of belts and hooks, this is often the method of choice. A ladder is also required to connect the drop installation to the house, so a ladder is always a part of an installer's necessary equipment. Figures 1-7 and 1-8 show some of the aerial lift truck uses.

The proper safety practices using ladders, climbers, and lift vehicles are covered in many company safety manuals. Many cable operators have safety manuals covering their methods and equipment. The Society of Cable Telecommunication Engineers covers climbing and ladder use in its installer training manuals, and lift truck manufacturers publish approved safe operation procedures.

1.234 Many innovative techniques were tried when direct burial cable was introduced to the industry. Farmers' deep bottom plows were tried. In some instances they worked quite well, but the cable placed in such plowed trenches was not deep enough, and stones were hard to find and remove. Equipment manufacturers soon introduced chain trenchers, which were welcomed by both the telephone and cable companies. Often a backfill "dozer" blade was mounted on the opposite end of the chain end for reversing the trenching direction, forcing the removed earth back into the trench after the cable was laid. Stones that could crush the cable had to be manually removed. When too many stones were found, sand and gravel were used to cover the cable before the trench was filled. Trenching is still widely used today, and for many types of soil it is the method of

Figure 1-7
Back-stringing strand
installation using a
bucket truck

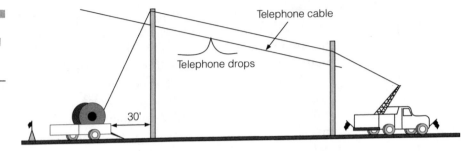

Figure 1-8
Back-stringing cable
installation using a
bucket truck

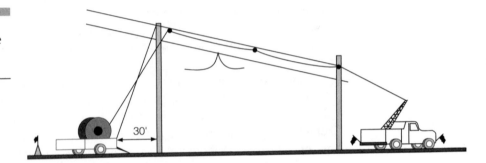

choice. Cable in conduit is often used today and is often faster because the backfill procedure is not critical.

As mentioned earlier, equipment manufacturers were quick to realize the potential advantages of opening a trench using the plow method. Today, two types of plows are used to place cable underground: the vibratory plow and the stationary plow. Both types require little or no ground restoration. The vibratory plow is inserted into the earth and when the plow blade moves forward, it vibrates back and forth, forcing the earth to open in a slit. Behind the blade is the cable chute, which slides the cable to the bottom of the slit. The earth simply folds over, covering the cable. This type of plow is usually faster then the stationary plow and requires less drawbar pull. Some vibration is transferred to the cable, but it isn't sufficient to damage coaxial, telephone, or electric cable.

The stationary plow operates in essentially the same manner as the vibratory plow but does not vibrate. Thus, no vibration is transferred to the cable. This type of cable plow is required for the installation of buried fiber-optic cable that cannot withstand the vibrations of the vibratory plow. The stationary plow blade is inserted into the earth and is pulled

Figure 1-9
Cable-plowing
techniques

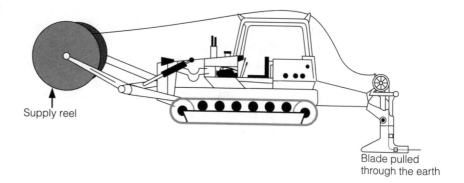

Supply reel

Blade pulled
through the earth

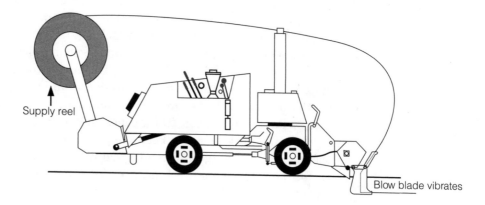

Supply reel

Blow blade vibrates

horizontally, forcing a slit opening where the cable is inserted through a chute following the blade. This type of plow requires more drawbar pull, which is provided by the drive engine and the tires or treads. Figure 1-9 shows these types of plows.

Another type of equipment used for urban cable system plants is the earth saw. This essentially is a huge circular saw with replaceable hardened teeth mounted on the edge. Rotating the saw with an engine can cut a 3- to 4-inch slit in granite, concrete, or bituminous concrete road surfaces. Moving the rotating blade in a horizontal motion can cut a slit in hard surfaces at several feet per minute. Cables installed in such trenches are usually in some kind of conduit. In most cases it is the integral cable conduit or steel pipe or PVC-type conduit.

For metropolitan-type installations, conduit is required by the Public Works Department. Restoration of the work area is specified by the municipal authorities. The cable conduit is often placed in soft sand at the bottom of the trench, which is then filled with concrete slurry to within a few inches of the top. The remainder is covered with the same material as the road or sidewalk surface. This is an expensive and noisy operation, but for urban/metropolitan installations, this is often the only method.

1.235 Metropolitan or city plant is much more expensive and complicated than the usual suburban, residential-type aerial and underground plants. Metropolitan areas often suffer from a variety of signal problems such as multipath and signal-level variations. These metropolitan areas usually cover large apartment complexes where television service had been provided by a master antenna system. Many apartment managers opted for cable television service when it became available. An apartment complex can make the usual cable television homes-per-mile requirement look good. Therefore, cable operators moved to the cities to provide service to these multiple dwelling units (MDUs).

Metropolitan plant is mostly an underground plant installed in either saw-cut trenches or in leased conduit space in the underground utility system. Cable operators using the underground utility plant have to use all the precautions and procedures as required by the conduit-leasing agency. Safety barricades, traffic cones, and barricade tape are required to mark off the work area and keep the public away. Ventilating blowers are used to purge underground vaults of any unsafe gases and provide clean air for the maintenance technicians. Proper personnel training should be provided by the cable operators to ensure safe working practices.

In many cases, the utility conduits pass under rivers and harbors in the transportation tunnels, or they cross bridges and overpasses. Access vaults are rarely placed on bridges or in tunnels; instead they are placed at each end. Thus, it is just the cables making the crossings. Cable amplifiers, couplers, and power supplies are installed in the vaults. Technicians making maintenance or troubleshooting checks have to gain access to the vaults to make the required tests. Cable service distribution is coupled from the vault system to the apartment buildings that contain the distribution amplifiers and subscriber taps.

An inter-building plant is either from top to bottom or bottom to top, with the cables placed in utility shafts connecting to each floor. At the various floors, a distribution panel is installed with drops connected to the apartments on each floor. In such urban areas, installation contractors often install systems in new apartment complexes being built. Once such systems are built, the cable operators need properly trained personnel to perform connects, disconnects, and maintenance tests.

1.3 Changes in CATV Regulations and Requirements

There have been many changes in governmental regulations. Most of the changes have been in the Federal Communications Commission (FCC) regulations and are a concern for many cable operators. Changes have occurred, too, in municipal or local regulations but have not been significant for cable operators. Both of these types of regulations will be discussed.

1.31 Federal Regulations

Federal regulations pertaining to cable television operations have increased in number and complexity over the years. Early start-up systems essentially had no type of regulations and at that time the commission could see no need for any controls. When cable systems improved and started offering imported television stations, the must-carry rules were formulated and are still with us.

1.311 When cable systems started to use the so-called mid-band—between television channels 6 and 7—for added services, the FCC imposed leakage regulations. The cable industry worked with the commission in developing the cumulative leak index (CLI) concept, which is in use today. These rules require leakage measurements of signal levels leaking from cable systems using a signal level meter and test antenna. The measurements are then used to calculate the signal in microvolts per meter and, by using a formula with the number of leaks measure, the CLI for the system can be calculated. This material is discussed in Chapter 2.

Another area of concern the FCC has for cable television systems is in the proof of performance tests and how and when the tests are made. As the cable television industry has progressed, the commission has become more concerned with the signal quality of the product. Systems now have to test for base band video signal quality. Proof of performance regulations and testing appear in Chapter 7.

1.312 Probably the most important regulation or standard adopted by the FCC is the recently accepted high-definition television (HDTV) standard. The work performed by the Grand Alliance and the FCC has produced the standard for HDTV. The Grand Alliance is made up of experts from AT&T, the General Instrument Corporation (GI), Zenith, Thomson, Philips, Sarnoff Labs, and the Massachusetts Institute of Technology. Its purpose is to recommend HDTV standards and methods of modulation. The FCC created the Advanced Television Systems Committee (ATSC) to complete the work of the Grand Alliance from an official standpoint. The ATSC–SI-issued Document A/56 outlines and addresses the issue for subscriber device and service for digital programming. Since its formation, Cable Labs in Colorado has been very active and supportive in working with ATSC. With all the work that the industry and the FCC has been doing to get HDTV and digital television up and running, many questions remain unanswered at this time. This means more work is to be done to resolve the must-carry questions, and the restrictions and controls applying to the cable television industry.

In December of 1996, the FCC gave the okay for broadcasters to begin testing the new digital television transmission. This is the standard for the broadcasting industry to follow for transmitting HDTV, video, and audio services to the public. Since new television receivers will be needed to receive HDTV signals, a time frame for conversion has been set. After 2006, all television broadcasters will be required to be transmitting digitally in either HDTV or NTSC video formats. Modulation for broadcasters will be 8-VSB, while cable systems will use the quadrature amplitude modulation (QAM) 16, 64, or 256 digital formats. These methods of television transmission will be discussed in Chapter 5.

1.312 From time to time, Congress has passed legislation specifying cable television regulations and will continue to do so. The cable television industry has evolved from the simple systems of the early 1950s and 1960s to a multi-billion-dollar telecommunications industry. Still, many small, simple systems that are long since debt-free continue to provide one-way multichannel service to subscribers. The normal course of events for the industry was to consolidate assets by mergers, thus forming several giant corporations. Such corporations activated the return path, expanded the signal bandwidth in the forward direction, installed fiber-optic plants, and activated pay-per-view services. Some even instituted telephone and data services to become all-service telecommunication providers. The FCC has to contend with these issues as the telecommunications industry grows.

The FCC issued its second report and order on December 31, 1998, that requires cable television systems to become a part of the Emergency Alert System (EAS). This order requires equipment to be installed in cable system hub/headends so that in times of emergency an alert tone and message crawl is placed across all cable channels. Broadcasters also have this capability and are responsible to provide the emergency alert to people receiving only off-air reception. Several manufacturers also offer equipment and installation guidelines to cable systems. This is a good and useful service to cable subscribers and could provide appropriate warnings of hurricanes, tornados, and disasters.

1.32 Local Regulations

Most of the local regulations are written into the licensing/franchising agreements between the municipality and the cable operator. These regulations or requirements, as the case may be, address local programming issues, cable channels allocated to community use, and the equipment to produce such programs. Regulations regarding the cable operators' right to work on the public ways are also part of the licensing agreement. Some more urban communities have more extensive rules and regulations concerning work along the public ways. The department of public works usually is the community administrator of these rules and regulations. The local telephone company and the electric company also fall under these same rules. Essentially, the rules and regulations are concerned with public safety and the protection of public property.

1.321 The cable television license or franchise for a municipality or community constitutes permission to work along the public highways and byways. This license or franchise has to be in place legally before construction can commence. Proper insurance coverage for public liability and property damage is a requirement due at the signing of the license. Often safety issues are written into such agreements. Also at the time of licensing, the pole attachment agreement with the pole owners, usually the telephone and power companies, has to be in place. Some communities themselves are the local pole owners. Communities that own their own poles are also usually the municipal electric power provider. They, in turn, lease space on the poles for telephone and cable television systems.

1.322 It is important to all parties concerned to resolve any issues or problems with the new cable television system. After construction of a system is completed, the pole owners usually inspect the plant to ascertain if it is in compliance with the agreement. In many cases, the local department of public works will want an inspection of the construction along the public streets. A representative of the cable television company and that of the town or county will walk or drive through the various streets to make sure that waste removal and restoration work has been completed in a satisfactory fashion. Once these inspections are completed, all parties are assured that construction and cleanup is over and settled to everyone's satisfaction.

1.33 Commercial and Consumer Electronic Development

One very important factor facing the cable television operator is the subscriber television set itself. Unfortunately, this device is beyond the control of the cable system and belongs in the consumer electronic manufacturing domain. As most of us in the cable television business know, relations with television set manufacturers have not been smooth throughout the years.

The organization representing the consumer electronics industry is the Consumer Electronic Industry Association (CEIA) or the Consumer Electronic Manufacturers Association (CEMA), which evolved from the Radio Manufacturers Association (RMA). The CEIA represents the manufacturers of consumer electronics equipment, most of which consists of home entertainment equipment.

One of the areas of contention between cable television operators and the consumer electronics industry is that of cable-ready television sets. With digital television just around the corner, there is a great need for CEMA, broadcasters, and cable television operators to settle their differences and get down to the business of resolving digital television issues. It is indeed in the best interests of the general public that straightforward and easy-to-integrate equipment be available to interface the television set and the cable system.

Cable systems have made the converter box available to subscribers, with some integral form of decoder/descrambler for the premium (extra-cost) channels. This box also has a companion remote control that enables the converter to select a cable channel and present it to the TV set on a fixed channel, usually television channel 3 or 4. Subscribers now have to keep their TV sets tuned to the converter channel, and the program channel is selected by the converter remote control. This renders the TV remote control almost useless, and having two remote controls is often confusing. Television sets that are advertised as cable-ready are able to tune to the usual cable television channels using the TV set's remote control. For non-premium services, this works best, but when premium services are desired, the converter-decoder set-top device and its accompanying remote is needed.

Some television receivers have what is called the smart remote, which in many instances is programmed to operate the converter as well as the VCR and the DVD. Now only one all-purpose remote is needed. Many manufacturers of VCRs also have programmable remote controls. Many home entertainment systems containing the TV set, VCR, surround sound system, and DVD system may have as many as five or six remote controls. This can result in confusion. A few examples of subscriber equipment connected to the cable system are shown in Figure 1-10.

1.331 Digital-formatted television is definitely coming—and soon. In the past several years, much experimentation has taken place with advanced high-performance television methods. The Japanese experimented with an analog system that showed significant improvement. However, with the

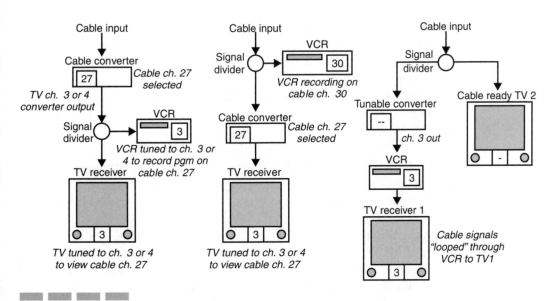

Figure 1-10
Examples of cable system–to–subscriber equipment

efforts of several manufacturers, the FCC and Cable Labs, as well as the Society of Motion Picture and Television Engineers, the digital method Motion Picture Experts Group (MPEG) II was approved and adopted. This MPEG II method carries the digital video signal in a compressed format along with CD-like multi-television sound (MTS) quality. The digital numbers forming the bit-stream activate pixels of the proper color on the display screen. In compressed form, only the changes in picture content are transmitted, resulting in a decrease in the overall bit rate required to provide a picture frame.

Using digital technology, the encryption or encoding needed for premium programming can be built into the digital information stream. For an aspect ratio of 4:3 (NTSC format) and with present video picture quality, the video signal in compressed format requires less channel bandwidth than the present 6-MHz television band. Thus, four equivalent television channels with present NTSC quality standards can be carried in the 6-MHz band with MPEG II video compression technology. This implies that each television broadcast station can gain three more equivalent channels, along with three times the advertising slots, by converting to the MPEG II digital compression standards.

Broadcast stations desiring to carry the HDTV standard signal form will require full use of the 6-MHz band for both video and sound. The HDTV standard uses a 16:9 aspect ratio and more pixels per picture, resulting in near-35-mm motion picture quality. Several broadcast stations in the three major networks are preparing their transmission facilities for digital signals. At this writing, it seems that the VSB-8 method of modulation will be used for over-the-air broadcasts and will be covered in Chapter 5. Cable systems are opting for QAM-16, 64, or 256 digital modulation for over-the-cable transmission.

1.332 As stated earlier, the initial trial into improved television transmission quality, during which several manufacturing companies experimented with several methods, was called advanced television (ATV). The resulting push for some sort of standard resulted in General Instrument Corporation's research on digital television technology. The FCC appointed a committee to investigate all of the proposed methods, and the acceptance of MPEG II digital video compression was the obvious choice. At first, the industry-sponsored Grand Alliance studied the ATV and HDTV proposals. An FCC-appointed committee was formed from this group. As mentioned before, the present MPEG II digital compression standard was approved. The FCC gave the green light to trials in 1998 and has allocated frequency bands to carry the new television digital format. The year 2006 is the target for full digital television broadcasting. Presently, several broadcasters are readying their transmission facilities for digital transmission trials and testing.

1.333 At present, digital television receivers are coming on the market. Most consumers seem to think some form of conversion box or converter will become available to convert the digitally transmitted format to standard NTSC format that is compatible with present-day TV sets. Although this seems to be a reasonable expectation, there appears to be no such equipment forthcoming. If broadcasters opt for a 4:3 aspect ratio digital television or an HDTV 16:9 aspect ratio digital television, it begs the question, does this mean two types of television sets will be required? And if 4:3 ratio pictures are shown on a 16:9-wide screen, will HDTV sets be accepted? Since there are no direct-view cathode ray display tubes in the 16:9 format, the type of display methods will be liquid crystal display (LCD) or plasma-type display screen. Some people believe that rear-screen projection type TV sets will be the most popular. Clearly, with the transmission of HDTV coming in the near future, television sets capable of receiving and displaying the new format should be ready approximately around the same time. Failure to do so will jeopardize the successful introduction of the service.

The 16:9 aspect ratio will probably take some getting used to in the living/family room setting. A comparison of 4:3 and 16:9 screens for nearly equivalent viewing areas is shown in Figure 1-11. For large screen areas, the 16:9 format may be too wide for many wall spaces. This could cause some problems with acceptance. The flat-panel display, mounted on a wall, will probably be the best method for HDTV wide-screen viewing.

At present, some VCRs use an internal digital video format, and the videodisc player also uses digitally recorded discs. Presentation to current NTSC TV sets would require the conversion to standard NTSC video/audio. In digital TV sets, the digital output signals from the digital tape

Figure 1-11

Television screen size for 4:3 and 16:9 aspect ratios

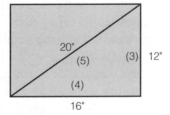

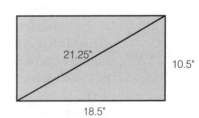

and disc machines connect directly to the TV set. Internal RF modulators in digital VCRs and disc players, as well as the RF channel selector in the TV set, should be circumvented so the digital signals will avoid any added noise.

Digital television promises improvement in picture and sound quality. Whether the television-viewing public will accept HDTV 16:9 aspect ratio in large-screen format, however, is anybody's guess.

Summary

In this chapter the step-by-step development of cable-type signal delivery systems is introduced. This development has produced several extremely wealthy large cable telecommunication companies in the United States and abroad. The basic need that gave the industry its start was poor television reception. Basically a coaxial cable delivery system containing an improved television signal originating from a hilltop, tower, or water tank was used to connect subscribers TV sets via a drop wire connected to the cable system. In many rural areas it was either have no television or connect to the cable system. Success of such cable television companies spawned the manufacturers to develop better cables, connectors, and devices that allowed cable systems to expand their business.

The basic characteristics of coaxial cable are discussed. Cable has a loss characteristic that is a function of signal frequency. The higher the frequency, the greater the cable loss will be. Special repeating amplifiers are used to increase the signal level when the high-frequency loss becomes critical. The unity gain basic building block concept states that when the high-frequency cable loss gets to the minimum point, an amplifier is used to increase the signal to make up for the loss. Improved amplifiers were developed that generated less noise and signal distortions, used less power, were more stable, and produced more signal gain.

Changes in plant design resulted in the tree-branch network concept. A coaxial cable trunk system using the unity gain building block theory made the tree-branch network concept work. Subscribers were connected to the signal feeder network containing a tap system. This feeder network is connected to the trunk network using bridging amplifiers. No subscriber taps are connected to the trunk network.

Construction of cable systems was either aerial or underground. An aerial plant is mounted on the utility poles along with the power and telephone systems. It is maintained by using a hydraulic lift truck, ladder, or belt and climber. An underground plant can be direct buried where soil conditions permit or placed in pipe or conduit. It requires chain trenchers or back hoes to open a trench for conduit installation or direct burial cables. Special earth plows have been developed that permit faster and more economical underground cable installation.

Regulation of cable systems installed on the public highways quickly came about as the industry developed. The FCC as well as individual states have laws with rules and regulations that the cable system industry has to follow.

As cable systems improved so did the consumer electronics industry with improved and larger television receivers. The greatest improvement was color television. The cable system has responded with more and better television programming to the viewing public.

Questions

1. List the problems associated with over-the-air television reception using an outside antenna.
2. What do the letters CATV stand for?
3. Describe what is known as "cable loss."
4. What did early cable systems do to carry the UHF channels?
5. What was the greatest improvement in coaxial cable?
6. Name the most significant improvement in cable system amplifiers.
7. List the main improvements in early cable systems.
8. Explain how lightning may affect a cable plant.
9. What two types of lightning protective devices are used in cable systems?
10. Explain how temperature or thermal problems are controlled in cable amplifiers.
11. Describe what is meant by the "unity gain building block."
12. What is the unit of measure for signal level in cable communication systems?
13. What cable parameter is most important for a cable power system?
14. Which cable type has the least loss, a large diameter cable or a small diameter type?
15. Name the advantages of jacketed cable.

16. Why are cable connectors so critical to a cable communication system?

17. What dictates whether a cable plant is aerial or underground?

18. Name the main or principal concern for cable plant construction.

19. List some of the advantages of cable plowing over trenching for underground cable plant.

20. List some of the problems associated with cable plowing over trenching.

Problems

Answers to odd-numbered problems appear in Appendix F.

1. For an impedance-matched system (for example, $R_{in} = R_{out}$), calculate the dB loss if the output voltage is 2 mV and the input is 200 mV.

2. Calculate the dBmV level for a voltage of 200 mV.

3. For a length of cable, if the input signal level is +3 dBmV and the output signal level is −23 dBmV, calculate the cable loss.

4. For the unity gain building block consisting of a length of cable followed by an amplifier, if the input signal level to the length of cable is +7 dBmV, what will the amplifier output level be?

5. For the cable circuit in problem 4, what gain will the amplifier need if the cable loss is 20 dB?

6. Calculate the dBmV signal level for a voltage of 1 volt?

Coaxial Cable Systems and Networks

Objectives

After learning the material in this chapter, the student will be able to

- Describe the coaxial cable networks that distribute programming to subscribers.
- Explain the bidirectional, or two-way, capabilities of a coaxial cable distribution system.
- Describe the operation and features of the electronic amplifiers.
- Explain the method of cable powering the amplifier cascade.
- Understand the various construction methods for both aerial and underground plant.
- Describe the methods used for protection and security of television programming.
- Understand the system testing and the necessary records and maintenance procedures.

2.1 Introduction

The basic idea of the cable television system was that the central antenna system would be shared among connected subscribers willing to pay a monthly fee. The headend, as it is still known, was the source of the signals making up the basic service. The headend was usually located on a hilltop, tall building, water tank, or steel tower located out of town, away from subscriber homes. The basic off-air received television signals were amplified, filtered, adjusted in level, and combined and delivered by the cable amplifier cascade to subscribers. This elementary topology was known as the tree-branch cable system. The main cable was the trunk system from the headend to the town where the cable branched down the streets to the subscribers' homes.

2.11 Tree-Branch System

The basic tree-branch system consisted of cable-amplifier sections feeding signals one way from the receiving headend tower site to subscribers. Many early cable operators were aware of possible two-way operation by splitting the service signal band between downstream and upstream service and installing upstream amplifiers. But that was basically wishful thinking, mainly because of system problems such as temperature stability, signal-level control, and amplifier noise and distortion. Early systems were one-way—even today many essentially still are.

2.111 The bread and butter of cable systems operation was, and in many cases still is, the delivery of many cable television channels containing a variety of program choices to subscribers. When more channels were offered and cable system costs increased, subscriber rates were subsequently increased. The concept of program tiering allowed subscribers a choice of programming packages such as news, sports, and movies. The source of these programs came about with the introduction of satellite technology. Presently the number of program channels available from various satellite systems is enormous.

The two problems cable system operators have in today's market are (1) adequate system bandwidth needed for an increase in channels and (2) making the decision as to what programming will sell best. To repeat, the main income stream of cable system operations is the delivery of television programming to subscribers. It was thought at one time that auxiliary services such as electric meter reading and alarm systems would support two-way cable operations, but the cost of the reverse system activation is a significant expense compared to projected revenues for the services.

2.112 Identifying the service area of a cable television system is a difficult problem. First and foremost is the fact that the areas of residential population are the criteria for successful subscriber penetration. Unfortunately, municipal authorities may impose requirements as part of licensing or franchising agreements that nonproductive areas be served. Also, future areas of residential expansion may not be readily identified, thus necessitating future plant expansions.

Most cable system operators avoided building cable plants in industrial and commercial areas. Thus, many cable operators have had to build more plants to provide information services to these areas. The location of the subscriber office area, the technical maintenance area, and the head-end location can also be factors in planning and specifying the service area. Essentially, the service area has to be specified before financing and construction plans can commence. Thus, as the saying goes, do the homework carefully and diligently. Technical people as well as managerial people should be involved in this decision-making process. Costs can be projected by estimating the miles of the plant, the feeder-to-trunk ratio, the number of program channels, the cost per mile of plant, and the head-end. Now the basic revenue can be projected, producing a cash-flow, or financial, pro forma to offset debt and support the venture. This process is complicated and has to be carefully completed before setting any construction schedule.

2.113 The proper plant design required to serve the more heavily populated area of many communities should not be so trunk restrictive that businesses will box themselves in by making it difficult to extend the distribution cables to new areas. In short, a system does not want to run out of trunk—that can restrict extending the feeder plant. Remember, it is the feeder or distribution plant that connects the cable system to subscribers. The trunk is considered as the transportation system bringing the service to the distribution plant.

The feeder/distribution plant makes the money, and the trunk cable plant is a necessary cost. The extension of a feeder plant adds more subscribers, thus improving the cash flow. Since the cost of extending a feeder plant is significantly less than the trunk plant costs, a careful analysis has to be made when extending a plant. Often a significant extension of the feeder plant means adding more trunk plants. This is shown in Figure 2-1, where feeder plant amplifier cascades are limited to two amplifiers in a cascade off the bridger trunk amplifier.

Adding more cable amplifiers to a system also means that system powering may have to be restructured with the addition of more power sup-

Figure 2-1
Trunk extension
required to extend
service area

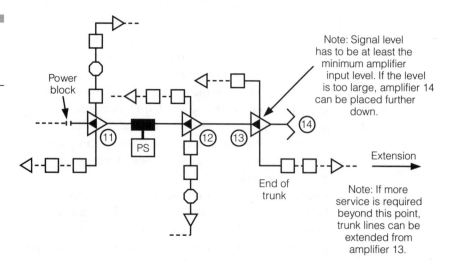

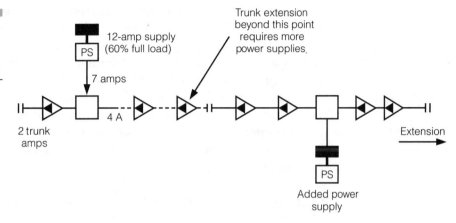

Figure 2-2
Added trunk requires
added system power
supplies

plies and re-segmenting the powering system. Re-powering such plant additions are shown in Figure 2-2.

2.12 Subscriber Equipment

The main receiving device of a cable television system is the subscriber's television receiver. These receivers can range from new to almost museum-piece TV sets. However, since TV sets have an average life of about 10 years, it is fairly safe to assume that most subscriber sets range from new to 10 years old. The median age range is estimated to be about four or five years old. Also, almost all are color TV sets. Since cable systems carry the service in standard NTSC VSB format, the television receives the signals and makes the channel selection by frequency-synthesized tuners, the same as if it were an off-air television station. Since cable system channels can be offset in channel frequencies in the HRC or IRC format, television sets made in the last 10 to 12 years have a switch often appearing on the rear cover or selected by the remote control labeled Cable 1 or Cable 2. This switch controls the action of the channel selector to tune to the desired channel settings.

The present-day television set is devoid of horizontal and sometimes vertical hold controls, due to vastly improved circuitry that locks the picture synchronization. Automatic chroma controls and contrast controls keep the pictures at the same color and brightness. The remote control system can usually address all picture quality controls, channel selection, and video source selection.

Integrated circuit technology, circuit board construction, and modular tuning, along with the remote control system, have reduced the cost of TV sets drastically. Still, the picture tube or the projection system determines the cost of the receiver. All of the electronic systems of a number of today's TV sets are valued at $40 to $50. The cabinet and screen/tube system results in most of the cost of a TV set. Since the subscriber's television set is what produces the cable system's picture, it is most important that cable operators make sure the cable installer makes the proper connections to the TV set and instructs and advises the subscriber in getting the most out of the service.

2.121   As stated earlier, the television set has to tune precisely to the cable channels for best signal and sound quality. The noise characteristics of the television set tuner are most critical. The noise figures of early TV sets were quite poor and the UHF portion even worse. As TV sets improved, the noise figures of the channel-selecting tuners did as well. The IF bandwidth characteristics improved also, which in turn gave better pictures with improvements in the quality of TV receiver design. The picture resolution improved from about 260 lines to about 380 to 400. With the arrival of HDTV, picture resolution will indeed improve again.

Since HDTV is be digital, the consumer electronic manufacturers are going to be busy designing, developing, and testing TV sets for this new service. The broadcasting companies are busy getting new studio and transmitting equipment capable of handling the digital signals, readying the public for HDTV service where no HDTV sets exist. The trials of off-air digital television services will be taking place soon. Then maybe the problem areas will be identified. As yet no signal strength parameters, signal-to-noise specifications, or tolerance-to-multipath reception information has resulted. In short, no one knows if digital television broadcast stations can be received with rabbit-ear antennas or will be back to roof-top antennas.

As far as the cable operator is concerned, questions remain. First, will cable headends receive digitally modulated signals from a tower-mounted antenna? Second, will cable systems convert some of the signals to other modulation schemes or to NTSC VSB format? Third, what will be the must-carry rules set forth by the FCC for broadcast stations transmitting multiple-channel digital signals? And the list seems to go on and on.

2.122   The channel converter either rented or purchased from the cable operator has traditionally been required to provide cable television chan-

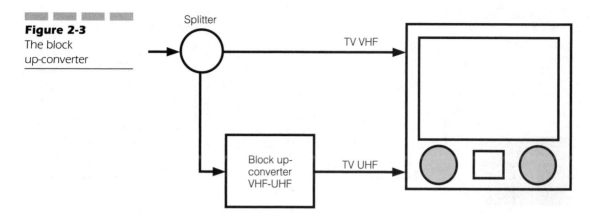

Figure 2-3
The block
up-converter

nels unable to be selected by standard TV sets. For some early cable systems that placed added programming channels in the so-called mid-band, a simple UHF up-converter was used. This device simply converted the mid-band channels that were spaced from the FM band to VHF channel 7 to the UHF band (470 to 890 MHz). This block converter, as it was known, connected between the cable and the TV set's UHF terminal, as shown in Figure 2-3.

Since there were 9 equivalent 6MHz television channels spaced in the mid-band, these 9 channels, in addition to the 12 VHF channels, resulted in what was known as the 21-channel cable system. At this point in time, available trunk and distribution cables had significant loss at UHF frequencies. Available amplifiers had upper frequency limits of 220 to 300 MHz.

The reception of UHF broadcast stations was carried on the cable systems by converting each one of them to one of these 9 mid-band cable channels at the headend. The block up-converter at the subscriber's television set would then convert them back from mid-band to UHF. Back in the 1970s the 21-channel cable system offered a decent selection of programming with better reception than most indoor or outdoor antennas. The success of 21-channel systems encouraged the expansion of cable operators to increase the upper frequency limit to 300 MHz. Thus, another 14 television channels were added, making the 35-channel system. This increase in channels beyond the frequency selection capabilities of the TV sets of the day made the tunable, or frequency-agile set top, converter necessary. The method of connecting this device between the cable and the television set is shown in Figure 2-4.

Figure 2-4
Tunable subscriber
converter

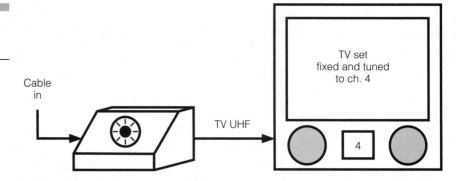

Converters of this type had to have good noise specifications, or low noise, simply because the cable signal has to go through the front end of the converter and then through the front end of the TV set, resulting in noise contributions from both sources. These first-generation tunable converters required the subscriber to select the desired program on the converter, which translated to a fixed channel (usually channel 3 or 4). This situation essentially defeated the use of the set's remote control. Thus, the first big problem with subscriber devices resulted.

The natural progression to increase the upper frequency limit beyond 300 MHz led to the evolution from 52 channels at 400 MHz to 62 channels at 450 MHz to current state-of-the-art systems offering more than 150 channels.

2.123 With the tremendous increase in cable system channel capacity, subscriber service selections became enormous. Many subscribers wanted only a few basic channels covering local news, weather, and sports. Therefore, cable operators tiered the program selections according to their subscribers' viewing habits. Since all cable system programs were on the system, it was natural to use a filter installed at the service drop to prevent the undesired program bands from being fed to the subscriber's home. In the industry this was referred to as negative trapping; that is, the unordered services were trapped out of the subscriber service drop.

With the arrival of the satellite system that made available premium (pay) channels to cable operators, the negative traps could prevent these channels getting into homes not subscribing to these channels. System audits are conducted by driving the aerial plant, looking for the presence or absence of the traps. The trap method can be defeated by a subscriber climbing a ladder to remove the trap. Determined cheaters replace the trap filter with a look-alike device to fool the cable operator. Hence, the pilfering of cable services started and program security became an issue.

Today cable operators have developed various programs to control the theft of cable services. From negative signal-trapping methods, the positive trap method was developed. This method engages an interfering carrier system where an interfering carrier signal was injected in the band of the signal to be protected. At the subscriber tap, a trap would filter out the interfering carrier, allowing the signal to enter the subscriber drop without the interfering signal so the subscriber could receive the premium program. This method meant that only premium (pay) program subscribers would need this trap. The ratio of premium subscribers to nonpremium subscribers was an important factor for making the economic decision for the best method. In short, a system that does not have many pay-program subscribers needs a greater number of negative traps than positive-type traps.

Remember that the programming choices are the stock and trade of the cable television industry. In today's market where the computer Internet connection is a desired service, many telephone system service providers have difficulty in delivering large data files because of the limiting transfer speeds or low bit rates. A cable system's bandwidth can allow for faster downstream bit rates, but upstream ordering capabilities are in many cases not available. For cable systems operating in the present computer environment, the expansion of two-way services is a must. Bidirectional plants with sufficient bandwidth are necessary for cable operators to become players in the telecommunications field.

2.2 The Cable Distribution Plant

The coaxial cable system design consists of the trunk and distribution of coaxial cable sections of cascaded amplifiers. As mentioned before, simpler elementary systems of earlier designs consisted of single cable-connected amplifier cascades. No differences were made between trunk and feeder cables. It was soon determined that the trunk cable cascade should be the signal transportation medium and should just supply the signal to the feeder cable system containing subscriber taps and distribution type (line extender) amplifiers. This method extended the signal the greatest physical distance from the headend and became known as the trunk-feeder system, or rather simply the trunk-branch method.

2.21 Trunk-Feeder Tree Method

The trunk-feeder method became the most accepted type of cable system design, known as the bread and butter of system architecture. System design and layout was often done manually on a drafting table, overlaying the design on a series of street maps. Often maps can be obtained from several possible sources, such as

- The local telephone or power company;
- A local land surveyor/civil engineer;
- The local town or city department of public utilities; and
- U.S. Coast and Geological Survey maps.

Many utility companies also offer their maps on either floppy disk or CD-ROM. Once maps are obtained, then strand mapping can commence.

Since maps may not contain any late changes, a measurement using a measuring wheel or a surveyor's transit and stadia rod to measure the distance between utility poles may be necessary. If utility poles are not going to be used, then a wheel measurement along the buried plant route should be made. For an aerial plant on utility poles, a pole attachment agreement with the pole owners should be initiated and in the works. Once strand mapping is complete, make-ready work can commence. Since the measuring wheel is a low-cost and simple means of establishing the cable system length, cable operators should keep several handy for measuring added plants as well as subscriber drop lengths. Such a wheel is shown in Figure 2-5.

When the measurements are completed and the distances are entered on the map, then the poles and the routes of the cable plant are known. At this time, two projects can be completed at the same time. The first is the make-ready survey with the pole owners and is often a three-party survey between the cable operator representative and representatives of the telephone and electric companies. Each pole that is to be used for the cable plant is examined to determine if there is proper space for the cable system. If any of the telephone or electric wires need to be moved to make room for the cable system, the cost is charged to the cable system. Most cable companies are well aware and are experienced in the use of leased space on the local utility poles. For underground or buried cable plants, only permission from the municipality is needed to perform the work. Responsibility for any damage to an existing buried plant such as telephone, power, water, sewage, and gas falls on the cable operator. It

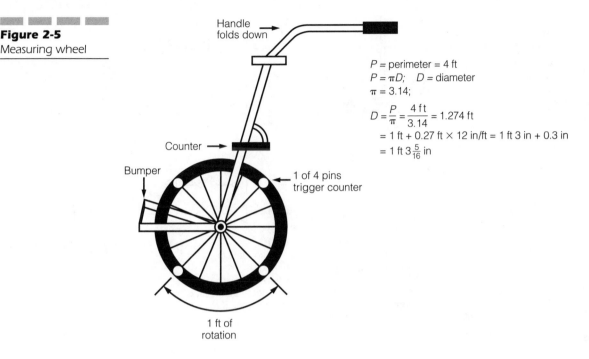

Figure 2-5
Measuring wheel

Handle folds down

P = perimeter = 4 ft
$P = \pi D;$ D = diameter
π = 3.14;

$D = \dfrac{P}{\pi} = \dfrac{4\,ft}{3.14} = 1.274\,ft$
$= 1\,ft + 0.27\,ft \times 12\,in/ft = 1\,ft\,3\,in + 0.3\,in$
$= 1\,ft\,3\tfrac{5}{16}\,in$

Counter

Bumper

1 of 4 pins
trigger counter

1 ft of
rotation

is usually the responsibility of each plant owner to mark out his or her respective plant on the request of any of the other companies.

2.211   Once the system routing has been established, the trunk layout can be designed. The trunk layout designed by the engineering department is based on the known area of system expansion. An example of a trunk layout is shown in Figure 2-6. From this point, a system design can be done manually using a pocket calculator or electronically using a computer. Several versions of software are available for personal computers and some are more involved and complicated than others. Computer-aided design and drafting programs have been around for some time. Most current ones can produce simultaneous trunk-feeder designs complete with tap values, port signal levels, amplifier input/output levels, and noise and distortion values, as well as a complete bill of materials. Because such software, computer, and plotting equipment are expensive, it is more cost-effective to have such work contracted out. Currently, several such design contractors exist throughout the country and most provide budget-conscious designs in a timely manner.

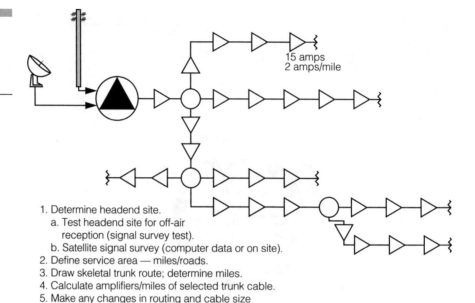

Figure 2-6
Trunk amplifier routing needed to cover the service area

15 amps
2 amps/mile

1. Determine headend site.
   a. Test headend site for off-air
      reception (signal survey test).
   b. Satellite signal survey (computer data or on site).
2. Define service area — miles/roads.
3. Draw skeletal trunk route; determine miles.
4. Calculate amplifiers/miles of selected trunk cable.
5. Make any changes in routing and cable size
   to cover plant layout.

2.212 The feeder or distribution system contains the subscriber taps, which are the part of the system that produces revenue. Thus, it is extremely important that it is also properly designed and constructed. Essentially, the feeder system is bridged off the trunk system through a bridging amplifier module installed in a trunk-amplifier housing. From this bridging amplifier, the distribution cable delivers a signal to a series of subscriber taps. When this signal level decreases sufficiently, a line extender or distribution amplifier is required to increase the signal level for another series of taps. Most wideband 750-MHz systems limit the number of distribution amplifier cascades to two amplifiers. Figure 2-7 shows a section of a distribution system, displaying what happens to signal levels along the distribution cable plant.

Notice that at the bridger output the signal has a 9-dB slope and, due to taps through loss and normal cable loss, the signal level at the last tap port nearly has a 9-dB tilt (reverse of slope). The higher the upper frequency limit becomes, the more severe the tilt. Thus, the insertion of an inline equalizer before the last tap will provide signal correction. If the system is to be extended beyond the last tap, however, a line extender amplifier with a proper equalizer will be required. Normally, most systems limit the number of line extender amplifiers to two in a cascade. This

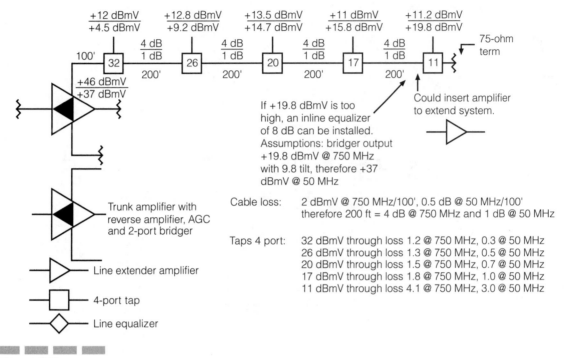

Figure 2-7
Example of a distribution section of a cable plant

results in three cable sections of five or six, four-port taps, servicing 65 or 70 homes.

As the upper-band limit is expanded from 750 MHz on up to 1 GHz, the spacing between amplifiers for both the trunk and feeder systems shrinks. Now more amplifiers per mile are required as well as more power supplies to feed them. The constant fear is that the buildup of noise and amplifier distortion products would limit the system length or reach. In some instances, it becomes possible to change system routing to control noise and distortion problems, but nothing can be done about the high power consumption of a wideband coaxial cable system plant.

2.213 Cost-effective system branching of the cable television distribution plant is very important. Maintaining proper signal level at the tap output port is of utmost importance to subscriber service quality for 750-MHz and higher systems. The approximate loss of RG-6 with 90 percent braid cable is 5.5 to 6 dB/100 feet. Therefore, a system should have a tap

Figure 2-8
Directional couplers
used to branch A, B,
and C

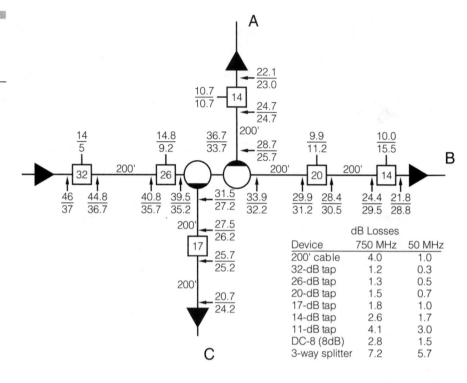

dB Losses

Device	750 MHz	50 MHz
200' cable	4.0	1.0
32-dB tap	1.2	0.3
26-dB tap	1.3	0.5
20-dB tap	1.5	0.7
17-dB tap	1.8	1.0
14-dB tap	2.6	1.7
11-dB tap	4.1	3.0
DC-8 (8dB)	2.8	1.5
3-way splitter	7.2	5.7

port level of at least +10 dBmV at the 750-MHz frequency. Proper drop cable shielding effectiveness is also necessary to control signal leakage and ingress.

You will often have several choices for branching within a service area design. If a manual design method is being used, the designer can ponder the various choices, taking signal level and cost factors into consideration. Signal splitters and directional couplers are the passive devices used for system street branching. Figures 2-8 and 2-9 illustrate two methods serving the same residential street layout. One method employs a balanced output splitter and the other one uses directional couplers. Figure 2-8, using two direction couplers, has two short branches with one four-port tap each and one long branch with two taps before a line extender is required. Figure 2-9, on the other hand, could use two taps in three directions, with the last tap a terminating tap before a line extender amplifier is required to extend the system.

Computer-aided design and drafting programs keep a running tabulation of signal levels. When this level approaches a predetermined level,

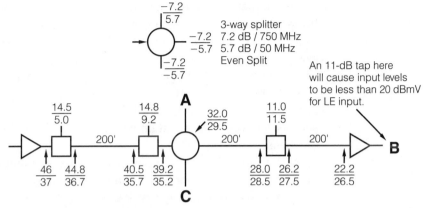

Figure 2-9
Balanced three-way splitter for branches A, B, and C

Note: All signal levels are +dBmV.
Top figure is @ 750 MHz.
Bottom figure is @ 50 MHz.
Legs A, B, and C are identical.

then an amplifier is needed at that point to continue the cable run. Where branching is required, the designer operating the CAD system can make the branching choices. When an amplifier is inserted, the program then annotates the system map with such parameters as input/output level, pad value, equalizer value, and carrier-to-noise and carrier-to-distortion calculations. These parameters can be very useful when compared to actual system measurements at initial proof of performance time. Trunk system splitting and coupling is equally important in determining the system network topology. A system designer should have a good general idea where any areas of future expansion might be proposed. Designing in some additional trunk during an initial build can often pay off down the road when a new area is developed. Only local people living in the area will know this information.

2.22 Cable System Electronic Equipment

Cable system electronics have come a long way from where they started, from vacuum tubes to transistors and now to integrated circuits (ICs). Today's cable amplifiers have more gain, higher output operating levels, improved noise and distortion figures, and less power consumption. Equally important is the high construction quality that gives the industry a superior, more reliable device.

The industry has also become more knowledgeable about the metallurgy involved in cable systems, resulting in corrosion-resistant and weather-tight amplifier housings. Modular construction of the amplifier sections has increased the flexibility and serviceability of the present-day cable amplifiers. Most cable amplifiers can be operated at more than one power supply voltage. Most systems operate at 60 v.a.c., more progressive systems operate at 90 v.a.c. Jumpers in the power supply section can be set to select the proper system voltage. Also, the stability of the amplifiers has been improved.

With proper thermal compensation circuits, along with automatic gain and slope systems, present-day amplifiers can provide many years of service. In many cases, a properly designed and constructed cable system will cause few problems, so service and technical personnel do not get enough service experience. When things do break down, there can be a lot of head scratching, as workers try to remember where to start and what to do.

2.221 The theory of cascaded amplifiers has been well known for many years. Radio communications people worked with short cascades to solve antenna preamplifier and strip amplifier signal distribution problems. Receiver designers used the theory to design multistage IF amplifiers. Telephone company engineers used cascaded repeater amplifiers in long haul or long trunk lines where telephone voice channels were multiplexed on a coaxial cable. Double sideband AM techniques were used where one voice channel occupied the upper sideband and another voice channel occupied the lower sideband of a carrier. The carriers were in the 10- to 100-KHz range, and groups of channels were formed according to CCITT international specifications. Telephone traffic was routed through coaxial cable–cascaded amplifier systems for much of the transcontinental runs. This technique was also used in microwave radio relay systems where the channel groups operating in the KHz bands were frequently converted to microwave carriers and were relayed by microwave hops. The microwave carriers could be converted to the lower-frequency ranges used on the coaxial cable system. Many of the early cable television engineers and technicians found a lot of information about cascaded amplifier techniques in the technical papers published by the telephone companies.

Cascaded amplifier fundamentals appear in Appendix E. Most people familiar with the theory use the mathematical formulas shown in Figure 2-10 and often work with them, designing system additions and extensions using a scientific pocket calculator.

As the upper frequency limits of cable systems increased, and performance specifications as well, the number of amplifiers in cascade had to

■■■ ■■■ ■■■ ■■■
Figure 2-10
Noise and distortion
formulas for a
cascade of n
amplifiers

Noise: $(C/N)_n = (C/N)_{n=1} - 10 \log n$

Composite second-
order distortion (CSOD): $(C/CSOD)_n = (C/CSOD)_{n=1} - 10 \log r$

Composite
triple beat (CTB): $(C/CTB)_n = (C/CTB)_{n=1} - 20 \log n$

Cross modulation
distortion (XMOD): $(C/XMOD)_n = (C/XMOD)_{n=1} - 20 \log n$

n = number of amplifiers in a cascade

be increased. That's because the cable loss increased as the service bandwidth increased, so more amplifiers were needed per mile. Also, as better carrier-to-noise ratios and carrier-to-distortion figures were needed, the number of amplifiers allowed in a cascade decreased. Such problems required many older cable systems to limit their bandwidth and the number of service channels. This problem is made clear in Example 2-1, where the upper frequency is 750 MHz with 110-channel loading. The sample calculations indicate that the amplifier cascade has to be limited to about five amplifiers if the specifications for distortion are to be adhered to. Clearly, long cascades are prohibited for wideband high-quality cable systems. Replacing the coaxial cable trunk cascades with fiber optical systems enables cable operators to provide 110 channels at 750 to 1000 MHz.

■■■ ■■■ ■■■ ■■■
Example 2-1
750-MHz,
110-channel loading

For a single amplifier with a noise figure of 10 dB and an input signal level of +15 dBmV, the carrier level can be found using the following formula:

$$(C/N)_{n=1} = 59 \text{ dBmV} + \text{signal input (dBmV)} - \text{noise figure}$$
$$= 59 + 15 - 10 = 64 \text{ dB}$$

To find n for an end of cascade C/N of 49 dB,

$$(C/N)_n = (C/N)_{n=1} - 10 \log n$$

$$49 = 64 - 10 \log n$$
$$-15 = 10 \log n \qquad \log n = 1.5$$
$$n = 32 \text{ amplifiers}$$

For a carrier to second-order distortion specification of 50 dB, the cascade number can be calculated by

$$(C/SOD)_n = (C/SOD)_{n=1} - 10 \log n$$

For a single amplifier C/SOD = 63 dB,
$$50 \text{ dB} = 63 \text{ dB} - 10 \log n$$
$$-10 \log n = -13$$
$$\log n = 1.3$$
$$n = \log^{-1} 1.3$$
$$n = 20$$

For a C/CTB and C/XMOD of 50 dB,

$$(C/CTB)_n = (C/CTB)_{n=1} - 20 \log n$$

For $(C/CTB)_{n=1} = 64$ dB,

$$50 \text{ dB} = 64 \text{ dB} - 20 \log n$$
$$-14 = -20 \log n$$
$$\log n = 0.7$$
$$n = 5$$

For $(C/XMOD)_{n=1} = 65$ dB,

$$50 \text{ dB} = 65 \text{ dB} - 20 \log n$$
$$-15 = -20 \log n$$
$$\log n = 0.75$$
$$n = 5$$

2.222 Trunk amplifiers then became available with a variety of characteristics, such as push-pull, power-doubling, and feed-forward techniques. Push-pull amplifiers provided the cable operators with an amplifier that had good second-order distortion characteristics. Adding automatic gain (AGC) and automatic slope control (ASC) provided cable operators with the workhorse of the industry. Several manufacturers offered such push-pull AGC/ASC amplifiers. Power-doubling techniques, as shown in Figure 2-11, gave the industry higher output levels needed to extend system reach.

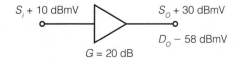

Figure 2-11

Power-doubling
amplifier

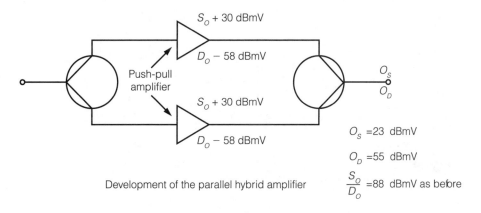

S_i + 10 dBmV S_O + 30 dBmV

D_O − 58 dBmV

G = 20 dB

$$\frac{S_O}{D_O} = +30 - (-58) = 88 \text{ dB} \qquad \frac{S_O}{D_O} = 88 \text{ dB}$$

Single push-pull-type amplifier

S_O + 30 dBmV

D_O − 58 dBmV

Push-pull
amplifier

S_O + 30 dBmV

D_O − 58 dBmV

O_S

O_D

O_S =23 dBmV

O_D =55 dBmV

$\dfrac{S_O}{D_O}$ =88 dBmV as before

Development of the parallel hybrid amplifier

S_O of parallel hybrid amplifier is 3 dB higher
with same $\frac{S_O}{D_O}$ ratio as a single amplifier.
A 3-dB improvement is equivalent to doubling
the power.

As bandwidth requirements and system length increased, the feed-forward amplifier was developed. This technique is shown in Figure 2-12. It does for third-order distortion what push-pull did for second-order distortion. Some cable operators extended cable cascades to nearly 45 amplifiers in a cascade for a 450-MHz 60-channel cable plant. Such systems operated at either an IRC or an HRC that allowed the third-order distortion levels to be relaxed by approximately 6 to 9 dB.

Essentially, any distortion products were moved to the edge of the television screens, which was less objectionable. Cable system headends had to be phase-locked to a reference comb generator, causing the cable operator a lot of expense. Each modulator and signal processor had to have phase-lock circuitry added so they could be phase-locked to the reference generator. Often phase locking at the headend, as well as using feed-forward

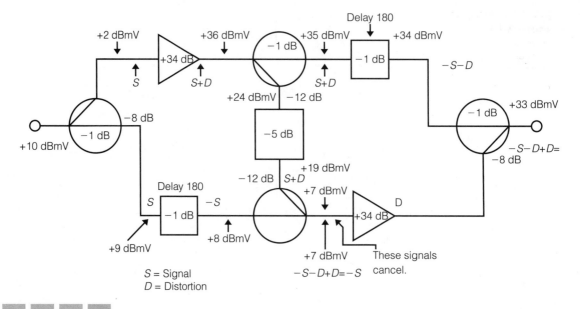

Figure 2-12
Operational example of a feed-forward amplifier

amplifiers, was necessary for cable operators to increase bandwidth and system reach.

2.223 Bridging and distribution amplifiers operate the feeder cable system containing the subscriber taps. It is imperative that the proper signal level at the tap output ports be maintained. Systems operating at 750 MHz with 110-channel loading will have difficulty feeding long subscriber drop lengths with adequate signal levels. Since the trunk amplifier's purpose is signal transportation, it is the job of bridging and distribution amplifiers to distribute the signal. The bridging amplifier module is installed in a trunk station housing and has the map symbol configuration as shown in Figure 2-13.

The distribution cable system consisting of subscriber taps, signal splitters, directional couplers, and line extender amplifiers connects to the bridging ports. A cascade of subscriber tap values in a descending order usually connects to a bridging port. When a signal split is needed, an amplifier may also be required to compensate for the tap through loss and/or splitter or directional coupler loss. System design computer programs keep track of the signal-level values and indicate when an ampli-

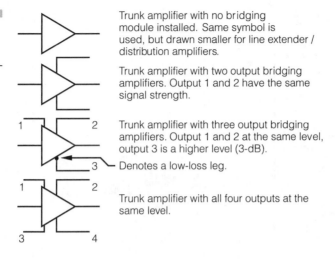

Figure 2-13
Map symbols for
trunk bridging

Trunk amplifier with no bridging
module installed. Same symbol is
used, but drawn smaller for line extender /
distribution amplifiers.

Trunk amplifier with two output bridging
amplifiers. Output 1 and 2 have the same
signal strength.

Trunk amplifier with three output bridging
amplifiers. Output 1 and 2 at the same level,
output 3 is a higher level (3-dB).

Denotes a low-loss leg.

Trunk amplifier with all four outputs at the
same level.

fier is required. Often, the exact amplifier location will be chosen by the
program operator. The amplifier location can then be placed on a more
convenient pole or pedestal, making maintenance and repair work easier
and less hazardous. It is the distribution plant that often provides the
greatest number of design choices available. In general, bridging and line
extender amplifiers operate at high gain and output levels; consequently,
the feeder cascades are limited to two amplifiers.

2.224 Systems using two-way services have reverse-amplifier and diplex
filters installed in the amplifier housings. These amplifiers operate in a
reverse direction to the main downstream cable system. A band of frequencies is allocated for the reverse system, and diplex filters are used
to separate the downstream and upstream frequency bands. The 5- to
40-MHz upstream band is below the 55- to 750-MHz downstream band
and is referred to as the subsplit reverse system. When the reverse band
is placed in the middle of the downstream or forward band, it is known
as a midsplit system. Placing the reverse band at the upper band edge
results in what is termed a high-split system. The most common of the
reverse bands is the subsplit system. This narrow band can be used for
approximately seven television channels, but the tightness of the system
determines how and where the noise buildup occurs in the reverse band
and the placement of the reverse carriers within the band.

The noise level in the active reverse band can be measured using
a spectrum analyzer. A typical spectrum analyzer display is shown in

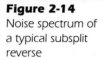

Figure 2-14
Noise spectrum of
a typical subsplit
reverse

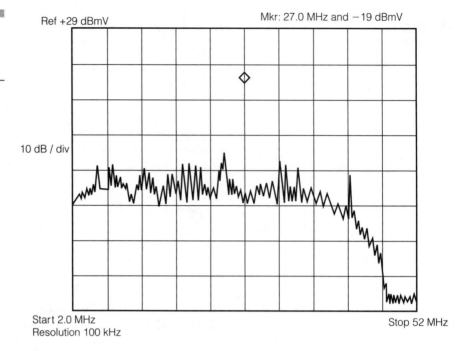

Figure 2-14. Before using the reverse system, the amplifiers and filters have to be installed. Since the cable loss is much less in the sub-band it determines whether an amplifier is needed at every station.

Design and projected performance calculations for each path in the reverse direction should be done. Again, many computer-aided design (CAD) programs contain sections on reverse system design. The reverse system should be activated on a leg-by-leg basis. When activating a reverse cable section, it will most likely be necessary to tighten or repair any leaky connectors, correct drop problems, and install reverse band stop filters in drops to reduce noise and signal ingress in the reverse band. In short, the system has to be reworked on a section-by-section basis. Now signals can be placed and balanced on the reverse system.

A spectrum analyzer measurement of the return signal can be measured at the headend. Several manufacturers provide instruments that are especially useful in balancing and monitoring the return signals. It is a constant maintenance problem to keep both the forward system from leaking signals and the return system from signal and noise ingress. Essentially, this means that all connector cables and device housings have to be tight and corrosion-free. Reverse system testing and maintenance topics will be studied in detail in Chapter 7.

Reverse or return systems have several areas of application in the cable television industry. One earlier application was to obtain programming from local schools, municipal meetings, and so on, which was done live and sent back to the hub/headend for downstream broadcasts to the subscribers. After local groups found how much work was involved in producing an hour's programming, interest decreased dramatically. Most subsplit return systems with two to four television channels available are found to be adequate for most community requirements.

Some of the more urban communities required a municipal institutional cable system called an "I" loop. Such systems could be simply a few sections of cable plant operating in both directions connecting several municipal locations together. One usual connecting point was the local cable operator's hub/headend. An example of such an institutional loop is shown in Figure 2-15. Such a system can be used to carry a full spectrum of signals in each direction carrying video (TV), voice, and/or data on modulated RF carriers. Switching functions at the hub/headend could broadcast any selected channel(s) over the forward system for subscriber viewing. This type of system was definitely a significant financial burden for the cable operator and, fortunately for most, few such systems were built.

Maintenance of such systems is essentially the same as for the forward system. For wideband forward and return systems, both a high- and low-pilot carrier is required at the sending ends. These carriers could be modulated with television signals and used as an ordinary channel, but they have to remain on the system used or unused as the pilot carrier reference needed to maintain proper system gain and slope.

2.23 System Construction Methods

Construction practices and construction equipment have definitely improved since the early days of cable television systems. The method for servicing an aerial plant used to be climbing the poles with a set of hooks and a belt. Most people learned to climb poles from earlier experiences working for either the telephone or electric company. Some learned to climb from a tour in the service, usually the U.S. Army Signal Corps. The Signal Corps adopted telephone company methods and specifications. Former Signal Corps veterans often became employed by cable television companies and learned the technical side during the early

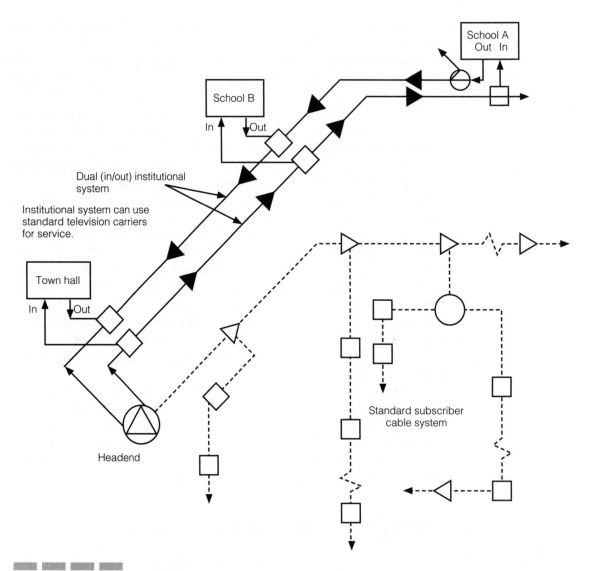

Figure 2-15
Institutional cable system overlaid on subscriber cable system

start-up years. Those that did not climb often used extension ladders against the poles.

Some innovative cable system technicians experimented with vehicle-mounted ladders. In a short time, the hydraulic lift truck became available and soon became affordable and practical for cable television system use. As suburban areas grew and became more populated, underground utilities, including cable television systems, became more attractive. Many system operators felt their plant was safely underground and at ground level for easy maintenance. Smaller and more affordable trenching machines made such installation financially viable.

2.231 Aerial construction methods have definitely undergone evolutionary changes. From a labor-intensive manual operation to modern hydraulically operated lift trucks with hydraulically operated drills, saws, and hoists, present-day construction practices have drastically improved. Some old-timers remember climbing the poles and boring through a pole with a bit brace and ship's auger. Nowadays a person is lifted up the pole in an aerial basket. The pole is drilled in about a minute and the through-bolt washers and nut are installed in another minute, enabling many more poles to be framed in a day. Such efficiency did not result in many cost reductions because of higher costs for equipment and more experienced operators, but a lot more plants could be built in a shorter time frame.

Pole-line hardware also improved over the years and more attention was paid to the effects of dissimilar metals on corrosion. Therefore, better materials were introduced. Usage of corrosion-resistant materials such as a Monel metal, stainless steel, and a variety of plastics produced a more reliable cable plant.

2.232 Underground cable plant construction in both methods and materials has changed more than aerial plant. First, the method of placing the plant underground has certainly progressed from "pick and shovel" methodology. Opening a trench has changed from a backhoe machine to a chain-type trencher that opens a narrow trench, spilling the removed earth into a neat row along one side of the trench. Usually stones are manually removed and a layer of gravel is placed in the bottom of the trench. A trench of this variety can be used with direct burial cable. Often PVC-rigid conduit is placed in the trench with an installed pull-rope to facilitate pulling the communication cable through at some later time. Semiflexible conduit is placed in the trench; some manufacturers provide the conduit with the cable already installed. Several conduits can often be placed in the same trench. Many trenchers also have an optional backfill blade that can plow the spill back into the trench. Trenching machines can provide trenches from very shallow to several feet deep. For most communication purposes, a depth of 24 to 30 inches is appropriate. A favorite method of placing cable underground is using a cable plow. As mentioned in Chapter 1, the two types of cable plows are the vibratory plow and the stationary plow.

Because an underground plant requires access to the active and passive devices, a variety of equipment enclosures are designed for underground plant servicing. These enclosures fall into two categories: above ground and underground. The aboveground enclosures are either metal or plastic pedestals, and the underground enclosures are either plastic or concrete vaults. The vaults are usually fitted with an appropriate cover that has to be either pried open with a special tool or unbolted with a special wrench. The aboveground pedestals have either a locking device or special tool. These measures prevent unauthorized tampering with the equipment. More progressive cable operators have electronic monitoring capabilities, called status monitoring, which can integrate an alarm system for vault or pedestal protection. Some of these pedestals and vaults are shown in Figure 2-16.

Figure 2-16
Types of pedestals
and vaults

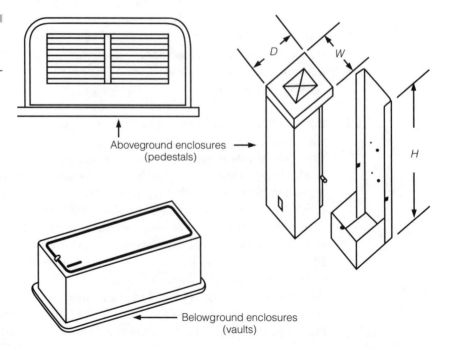

Aboveground enclosures →
(pedestals)

Belowground enclosures
(vaults)

Underground cable plants have to be mapped and properly marked so that service and maintenance can be done at a later time. Unlike an aerial plant, an underground or buried plant, as it is often called, cannot be seen. Therefore, the exact location of cables has to be known and appropriately mapped. Most of the problems with a buried plant result from the careless cutting and digging by other underground utilities. The installation of traffic signs often causes service interruptions for the cable, power, and telephone operators. Thus, various measures are employed to prevent damage to an underground cable plant. The installation of a yellow or orange plastic warning tape placed over the cable a few inches below the ground's surface offers some protection. Warning decals placed on the pedestals can also alert others that facilities are buried nearby.

Cable operators with an underground cable plant should have some type of cable time domain reflectometer (TDR) available to find the distance from one point to the fault. Making a measurement from each end of a cable with a fault (a cut or a crush) will give a close location of the fault. Also, cable operators need some type of cable location instrument, several types of which are commercially available. They place an electrical signal on the cable that is detected above ground by a receiver. In most cases, the transmitted signal is placed on the cable's metal outer sheath at one end of the cable run and a person with the receiver walks along the cable run, detecting the path of the cable by marking it with either spray paint or garden lime.

Once the distance to the fault has been determined and the path of the cable is found, the close location of the fault is identified. Following this procedure minimizes excessive digging and makes repair of the problem much faster. This method is illustrated in Figure 2-17. Underground plants are becoming more and more attractive to cable operators simply for the reasons of maintenance and safety from vehicle accidents.

2.233 Cable connectors have also improved over the years. One problem with connectors is what is known as a pull out, in which the cable is simply pulled out of the connector, causing a loss of signal and cable power. This is the most catastrophic problem with a connector.

Another problem is corrosion and connector deterioration over time. Systems with such problems will be faced with accompanying outages and subscriber complaints. More attention is consequently being paid to the metallurgy involved with cable plants and the problems of mixing various metals. Certain metals when in contact with one another in a moist and corrosive climate (such as air pollution or acid rain) can cause rapid deterioration of the plant. Also, certain metals used together in housing and

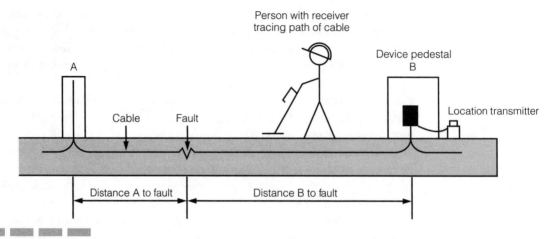

Figure 2-17
Location of buried cable path

fastenings can cause threads to seize. Different metals have different thermal characteristics that can cause expansion and contraction problems over seasonal temperature variations. Fortunately, several excellent companies provide high-quality connectors to the cable television industry.

Another consideration cable operators have to make is the ease of connectorizing the system, which means the ease of preparing the cable and installing the connectors. Proper tooling is a necessary requirement. Present-day connectors require that some of the dielectric have to be removed so the internal RFI sleeve can fit under the aluminum sheath. This operation is known as coring and can be done with a manual corer or one fitted on an electric drill. The drills used by cable splicers when connectorizing a plant can be powered by electricity (portable), by compressed air, or by hydraulic means, all of which speed up the connectorizing process. Once a connector manufacturer is chosen, the application's engineering department and/or customer service department will usually be most helpful to the cable operator, providing installation and tooling information. Most splicing contractors are familiar with the available connectors and the installation methods required. Proper installation is most important in keeping the power flowing and the signal inside the cable.

Coaxial cable systems have to conform to the FCC's required signal leakage specifications, testing the plant several times a year for signal leakage. Connectors that prevent signal leakage also shield the plant against extraneous signal ingress and noise ingress. Improper connectors do not give a good impedance match, which can result in standing waves and signal echoes, which result in ghosting of the television picture. Good connectors properly installed will provide good cable plant operation over the years. Some of the cable preparation tools appear in Figure 1-6 in Chapter 1.

2.234 Proper system grounding and bonding is extremely important to protect cable systems from the effects of lightning and power system surges. For aerial plants, the practice is to bond at the first and every 10th pole, as well as the last pole in a cable run. Some pole attachment agreements call for different grounding and bonding practices that must be followed. In most cases, all power supply locations for either aerial or underground plants must have a separate ground rod. Ground wires on utility poles often require the installation of protective pole molding to protect the ground connections. Special bonding clamps are used to bond copper wire, usually number 6 (AWG) soft-drawn copper wire, to the galvanized steel messenger strand supporting the aerial cable system. These clamps are usually plated brass. Some aerial bonding methods are shown in Figure 2-18.

Grounding and bonding at the headend and/or hub distribution points is crucial for protecting sensitive electronic equipment from lightning and power surges. Presently available are surge protectors that can be wired into the power entrance panels of hub/headend facilities. Such devices can protect all the equipment inside the building that is connected to the inside power distribution system. Emergency standby power systems of the external engine-powered variety also have to be properly grounded and bonded to the central grounded electrode. The main ground electrode is connected to a system of driven ground rods and the power entrance ground. A hub/headend ground and bonding plan is shown in Figure 2-19.

2.24 Subscriber Channel Converters

Subscriber- or customer-used equipment provided by the cable operator is essential to the delivery of a high-quality product. This equipment should be user-friendly and attractive in the television viewing system because it appears near or on top of the subscriber's TV set. Unfortunately, the use of a set-top cable television converter makes the subscriber's TV set remote control quite useless because channel selection is done by the set-top cable converter, as discussed earlier. This also makes VCR recording for the subscriber quite difficult and compounds the problem of the VCR remote control. Signal security in most cable systems is performed

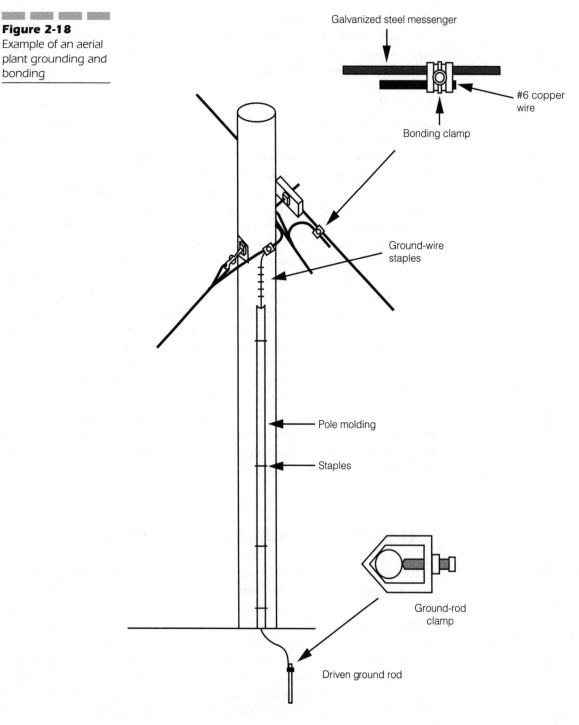

Figure 2-18

Example of an aerial plant grounding and bonding

Galvanized steel messenger

#6 copper wire

Bonding clamp

Ground-wire staples

Pole molding

Staples

Ground-rod clamp

Driven ground rod

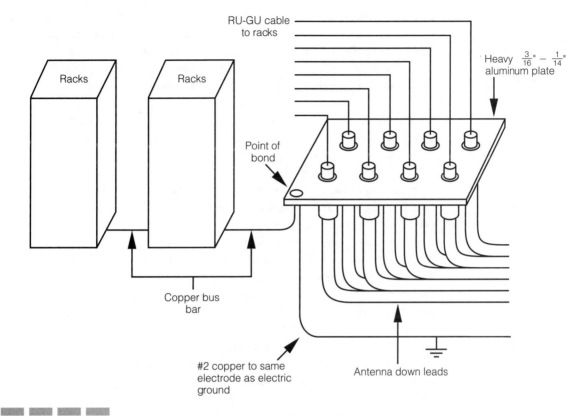

RU-GU cable to racks

Heavy $\frac{3}{16}$" – $\frac{1}{14}$" aluminum plate

Racks

Racks

Point of bond

Copper bus bar

#2 copper to same electrode as electric ground

Antenna down leads

Figure 2-19
Example of a headend ground/bond system

within the cable converter where the premium channels are decoded or descrambled.

2.241   The set-top converter has evolved over the years from the simple mid-band to the super-band channel selector. Most cable converters still perform channel selection and operate with their own remote control. Decoding and descrambling takes place within the converter for the premium pay channels.

Ordering premium-channel programming is still mostly done by a phone call to the cable operator's subscriber office. In many instances, a cable operator uses what is known as addressable converters. When a subscriber calls the cable office and orders a premium channel or a pay-per-view program, the converter is authorized to decode the program by a downstream data channel. This authorization activates the converter to decode the premium programming ordered. Cable systems with active reverse systems can place the ordering requests through the bidirectional set-top converter to the hub point. The subscriber can now order a particular pay-per-view program by using the set-top remote control to place the order. This type of television service is enough to make us all couch potatoes.

2.242 Prior to such sophisticated converters, signal security was provided by what is known as signal trapping. The unsubscribed premium program channels were trapped out (prevented) of the subscriber's drop cable. This was done at the tap port feeding the subscriber's drop cable.

Trapping devices appear as one of two types. A device that prevents the premium signal from entering the subscriber's drop is essentially a narrowband stop filter, known as a negative trap. The other trapping device is known as a positive trap. At the hub/headend, an interfering signal is placed on the premium channel that renders the channel unwatchable. The positive trap removes the interfering signal from the premium channel, now rendering it watchable. All subscribers not wishing to have the premium channel will need a negative trap, while a subscriber wanting the premium channel will need a positive trap. Systems that have a negative trap scheme with few subscribers taking the premium service will need a lot of negative traps. For such systems, a positive trap scheme will be more cost-effective because the few premium-program takers will need the interfering signal removal traps.

Since these traps appear at the tap port, a system accounting (audit) of subscribers taking or not taking the premium service can be made by a simple drive-by for observing the aerial traps. Underground systems where the taps are in pedestals should be locked to prevent tampering with the traps. Removing a negative trap allows the premium program to be seen. A positive trap of the proper type is needed to remove the interfering signal for the premium channel to be seen. Therefore, signal security is better using the positive trap method since an appropriate device in the drop has to be in place in order to receive the premium signal. Unfortunately, unscrupulous people sell devices that are usually inserted unseen within the TV set to those people who wish to steal the service. At the tap port, some people use dummy negative trapping filters that look like a trap but just pass the signal through.

2.243 Scrambling or encrypting a television signal, rendering it unwatchable, is indeed a better and more secure method of preventing stealing or unauthorized signal use. This is a more elaborate and expensive method, but many cable operators feel it is worth the expense.

Several methods are used to scramble or encode the premium channels. One method is to remove the horizontal synchronizing signal, commonly known as sync suppression. This type of scrambling is, of course, analog and works well with the standard NTSC video format. Inverting the horizontal synchronizing signal with some form of alternating pattern offers a higher level of scrambling. Such a method when working in conjunction with the subscriber addressable converter is still being used by many cable operators. With the arrival of digital television, digital encryption techniques will certainly improve signal security to a very high level. This apparently is what the cable television industry has been waiting for.

2.244 Addressable converters, or as they are known, addressable set tops, do two things for the cable operator. The most important consideration is that subscribers can order program services over the telephone, and the cable operator can then download the authorization codes to the set-top converter. Thus, the converter is activated to automatically make the changes and descramble any of the selected premium services. Addressability enhances signal security since the set top has its own address code where its location address is known. Now if this box is activated and stolen, and the service is not paid for, the box can be rendered inactive and hence useless.

2.245 Interactive converters are those set tops that can communicate directly with the hub/headend. Such converters make pay-per-view channels quite easy to be seen by any of the subscribers, and services can be ordered directly through the remote control. These converters can work the video on demand (VOD) systems some cable operators presently have in place. Converters operating in this manner must have a reverse path of communications back to the main signal distribution point. This reverse or return path usually operates in the sub-band (subsplit reverse) portion of the cable system. A relatively small number of cable systems have the return path in the mid-band or high-band that actually contains a wider return band, offering more equivalent upstream television channels. Interactive converters can be traced if stolen and detected in the wrong part of a cable system.

2.246 Connecting subscriber equipment to the cable system probably causes the most trouble for the subscriber. It is basically the set-top converter that causes the TV set and VCR remote control to be ineffective in selecting television channels. This problem causes many complaints from subscribers with cable service. The use of switches to change or reconfigure the subscriber's TV-VCR hookups may help the situation. Also, smart remotes (programmable remote controls) can be programmed to operate all of the subscriber devices.

This problem really resulted from the television manufacturer's inability to address the cable television industry's specifications. A possible solution would be for the TV set to operate at the intermediate frequency provided by the set-top box and the VCR. This subject has been under much discussion between cable operators and the consumer electronic manufacturers with the Cable Labs and the FCC advising at times. Several subscriber equipment connections are shown in Figure 2-20.

The National Cable Television Association (NCTA) released a publication addressing the interconnections between the cable converter, the TV, and the VCR. This publication covered a large variety of interconnections using A-B switches, splitters and couplers, multiple converters, VCRs,

Figure 2-20

Adding another converter, TV set, and VCR can make for a complicated home distributed system

Configuration A: TV Set 1 and VCR views the unscrambled channel tuned by the converter. TV Set 2 simply views any clear channel, if TV Set 2 is cable ready.

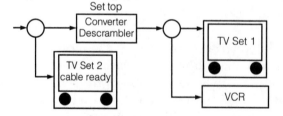

Configuration B: TV Set 1 can view any clear cable channel when A/B switch is in position A. VCR can be recording a pay TV channel selected by the converter. For the TV set to view the pay channel, switch to position B and tune the set to the converter output channel.

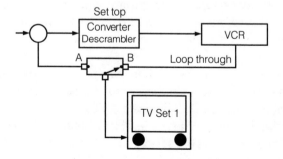

and TV sets. Many homes subscribing to a cable service that are supplemented by a direct satellite broadcast system (DBS) as well as a VCR can have an elaborate switching distribution system.

A word of caution: When a subscriber has an off-air rooftop antenna connected to the switching coupling network, care should be taken to prevent the wideband cable system from being connected to the antenna. This could cause some serious interference problems with many homes in the neighborhood as well as the cable operator. Remote as it may seem, it can happen and it has happened before. Such signals can disrupt air traffic communications as well as some of the law enforcement and safety mobile radio services. Another word of caution is to make sure switching and coupling devices are appropriate for cable television use. Some available devices are made for base-band video and are not appropriate for higher RF frequencies.

Probably one of the most important considerations with subscriber-connected equipment is the connectors themselves. Some of these connectors are used by the subscriber and are obtained at local electronic outlets, hardware stores, and homeowners' outlet stores. Unapproved tools and techniques are often used and hence the connectors are improperly installed. This is beyond the control of the cable operator, who only learns about it through resulting service calls reporting poor signal problems by the subscriber.

The connector used to connect subscriber equipment is the "F"-type connector, which had its infancy in the MATV industry. This connector evolved from a rudimentary connector with a separate 1/8-inch crimp ring to a single-piece, weather-tight connector that requires proper stripping tooling, and a more precise crimp tool. Several manufacturers supply versions of these superior "F"-type connectors to the industry. Most cable operators use this type of connector and have instructed their installer personnel in the proper installation procedures. However, a lot of the old connectors are still out there in many cable systems.

Changing these connectors is an expensive and labor-intensive procedure and is usually done during a system upgrade. These poor, early "F" connectors cause signal leakage, ingress, and noise problems that interfere with the use of an active return signal plant. When a cable system is planning a system upgrade requiring an activated return path, changing all the subscriber "F" connectors on an area-by-area basis is necessary in order to control the noise and signal ingress in the return path. Often subscribers not using the return path will have a filter trap at the return band frequencies, which will prevent signals or noise entering the return path from this location.

In many instances, cable systems discover signal leakage from a subscriber address during the usual periodic leakage testing required under the FCC rules. The usual procedure is to disconnect the subscriber drop and check the signal leakage level. If it goes away, then the subscriber will have to be contacted so an appointment can be made to make the necessary corrections. Since all of these costs have to be born by the cable operator, many cable systems avoid return path activation. An example of a current "F"-type connector and the accompanying cable preparation is shown in Figure 2-21. This is indeed an improved connector and can be used with single- and multi-shielded drop cables.

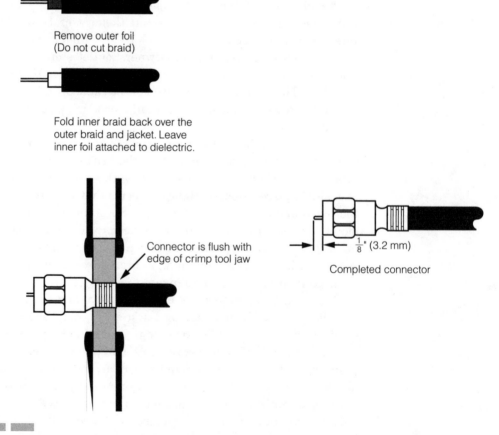

Figure 2-21
Example of a currently used "F"-type connector stripping and crimping

2.3 System Testing/ Proof of Performance

Cable television systems have to be completely tested on an annual basis for proper system performance parameters as required by the FCC under Part 76, Title 47 of the Federal Commission Rules of 1990. The first test period should be when the system is first turned on after construction is completed. A system operator should insist that the new system pass all FCC tests satisfactorily before final acceptance and the contractors are paid. In short, it should be a part of the construction contract in the final checkout and acceptance.

For the first three years, annual proof of performance tests are conducted on an RF basis, after which tests are performed on base-band signals. RF tests take place twice yearly and base-band tests are every three years. Systems with fewer than 1,000 subscribers are still exempt from proof of performance testing. Smaller systems are similar to MATV systems. Systems with 1,000 to 12,500 subscribers have to be tested at six widely separated test points, where at least one-third must be set at the most distant point from the central signal source. For systems greater than 12,500 subs, another test point needs to be added for every additional 12,500 subscribers.

Each system has to have in its company technical records a copy of the FCC rules and regulations Part 76, Title 47. Test records have to be kept on file for a five-year period and are subject to inspection at any time by a representative of the commission. The test results also must include the test procedures, calibration data and dates for the test instruments used, and the qualifications of the persons conducting the tests. The test points are specified at the subscriber terminal, which in most systems is the converter output terminal. Since most converters are pre-tested, tests can be made at the end of a 100-foot drop from the system tap and can be corrected by the converter performance specifications.

2.31 Initial Turn-On and Performance

Initial system test results are important in setting the system performance standards. In most instances, a newly tested system doesn't perform better than at turn-on time because system component and cable aging cause gradual system deterioration. A properly maintained and operated system periodically tests and services a plant to maintain proper signal quality. This, of course, is just good business, giving quality service to customers.

The turn-on sequence of events is important when performing initial tests. The procedure is to turn on plant distribution power, correcting any open- or short-circuit problems. Next, the pilot signals are injected into the system, and amplifier modules and power modules are installed in the trunk housings. At this time, each amplifier is roughly balanced on both the high and low pilots. Following this operation, a leakage test of the trunk runs is made so any further balancing will not compensate for a poorly impedance-matched plant. At the initial leakage test phase, the whole system can be swept using a sweep generator receiver instrument. This can be done on a system segment or a main trunk branch at a time. Last, the whole signal spectrum can be turned on and the system can undergo the final balance and leakage test. The results should be kept in the system file for future reference if any troubles should occur.

Often the costs of final testing are the responsibility of the prime contractor when the results prove to the system operators that the system is complete and operational according to the construction contract. Lending institutions holding the loan on the system may require the presence of either a consulting engineer as an observer of the final testing program or an independent testing company to perform the test to the satisfaction of all concerned parties.

2.32 End-Level Testing

The next most important test program is the periodic signal-level measurements of end points in the system. Unfortunately, many systems rely on the sporadic signal-level measurements made by drop installers or maintenance technicians making trouble calls. The measurement of system end levels on all cable channels can give a lot of information about how the system is operating. Seasonal variations of a signal level can indicate how well the thermal compensation and AGC/ASC systems are working. An analysis of this data either by a manual method or a computer program can indicate areas of the plant that need some maintenance long before the phones start ringing with trouble calls. Inexpensive graphics software can be used to present spectrum plots at the end points of the system.

For an aerial plant, a drop can be made at a tap fed from the last line extender and stapled down the pole. On the end of the cable, a connector installed with an F-81 barrel can provide a convenient point for a technician to connect a signal-level meter lead. The whole drop test point can be hidden behind pole molding. At the end of the test, the technician can remove the instrument lead and install a 75-ohm terminator to the test drop. Many instruments can provide the whole spectrum on an LCD display, and many can store data in a solid-state memory. End-level tests can be made either on a monthly or bimonthly basis, or at least on a quarterly plan, so data can be accumulated over a significant temperature range.

2.33 System Signal Leakage Testing Requirements

Another routine test that cable systems must perform is the signal leakage test required by the FCC. This test has to be done on a quarterly basis and the results must be kept in the company technical files. The cumulative leakage index (CLI) parameter will be computed for a system at the completion of the testing period. Any problems with signal leakage should be expeditiously repaired, with documentation on the cause, the fix, the date, and the personnel. The methods and technical specifications will be discussed in Chapter 7.

Two basic test procedures can be used to obtain the data for the CLI. One is the fly-over method, requiring the use of an airplane flying over the system at a fixed altitude. Monitoring equipment onboard the aircraft will record the signals leaking from the cable plant. Once the signal levels and the location are known, the problem areas can be easily identified. The use of the satellite-based global positioning system (GPS) can also be a big help in plotting the location of a leaking plant. The second method employs vehicle-mounted antennas used to locate problem areas. Once these areas are found, technicians using probe antennas can pinpoint the faulty connectors and loose amplifier housing. Many cable systems use this method. Technicians doing installation and maintenance tasks can monitor the plant at the same time. Keeping a plant free of leakage will also keep it free of signal noise and ingress.

2.34 FCC Leakage Specifications

The leakage specification for cable television systems depends on the frequency band of interest, the leakage intensity in microvolts per meter, and the measuring distance from cable plant as summarized in Table 2-1. The

Table 2-1

FCC Minimum
Leakage
Parameters

Frequency Band (in MHz)	Leakage Level (in μvolt/meter)	Distance from Plant (in feet)
Up to and including 54 MHz	15	100
Over 54 MHz, up to and including 216 MHz	20	10
Over 216 MHz	15	100

test antenna is specified as a dipole. However, a whip antenna mounted on a roving vehicle is often used to locate leaks, which are then measured with the dipole. The detecting receiver is either a signal-level meter of sufficient sensitivity or a special leakage-detecting receiver. Several instrument manufacturers make a combination test transmitter/receiver pair, and cable operators often have several receivers for ongoing leakage monitoring.

Example 2-2

If television channel 6 is used as a test frequency and the leakage measured is at the 20-μV limit, the following formula can then relate this level to a voltage value:

$$E = 0.0207F(V_r),$$

where $E = 20\ \mu V/m$, F is measured in MHz, and V_r is in μV.

$$20 = 0.0207(83.25\ \text{MHz})\,(V_r\,\mu V)$$

$$V_r\,\mu V = \frac{20}{0.0207\,(83.25)} = 11.62\ \mu V$$

This voltage value in dBmV

$$= -20\log\frac{1000\ \mu V}{11.62\ \mu V} = 20\log 86.06 = -38.7\ \text{dBmV}$$

If the signal-level meter cannot accurately measure signals, a small calibrated preamplifier can be used.

The dipole antenna should be tuned to the test frequency using the following method:

$$1\ \text{wavelength (ft)} = \frac{984}{F\ (\text{MHz})}\quad\text{and}\quad \text{dipole} = \frac{1\ \text{wavelength}}{2}$$

Each rod of the dipole will be 1/4 wavelength.

$$\text{Each rod length in feet} = \frac{984}{4F} = \frac{246}{83.25} = 2.95 \text{ ft}$$

and in inches,

$$2.95 \times 12 = 35.4 \text{ in} = 35 \text{ 1/2 in}$$

This method is illustrated in Figures 2-22a and 2-22b.

Now that the measured leak level in dBmV can be related to the specified leak in microvolts per meter, a sample calculation of CLI will follow. The method of testing is illustrated in the following example. (A sample of recorded data at television channel 6 is shown in Table 2-2.)

Example 2-3

Suppose this cable system has 60 miles of total plant and the preceding data was taken from a 20-mile section. The CLI can now be calculated.

At a measured level of -30 dBmV corrected by taking any preamplifier gain into account,

$$\frac{E\mu V}{m} = 0.0207(83.25)(V_r \,\mu V)$$

-30 dBmV corresponds to

$$-30 \text{ dBmV} = 20 \log \frac{V\mu V}{1000 \,\mu V}$$

$$1.5 = \log \frac{V\mu V}{1000 \,\mu V}$$

$$\frac{V\mu V}{1000 \,\mu V} = \log^{-1}(-1.5) = 0.0316$$

$$V\mu V = 31.6 \,\mu V$$

$$E\mu V/m = (0.0207)(83.25)(31.6) = 1.72 \times 31.6 = 54.5 \,\mu V/m$$

$$\text{CLI} = 10 \log \frac{60}{20} \times 134{,}512 = 10 \log (3 \times 134{,}512)$$

$$= 10 \log (403{,}536) = 10 \,(5.61) = 56.1$$

The upper limit of acceptable CLI = 64. Therefore, this system passes the test.

Figure 2-22a

Television channel 6
dipole test antenna

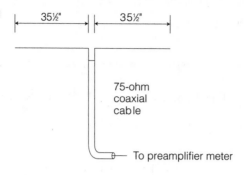

35½"   35½"

75-ohm
coaxial
cable

To preamplifier meter

Figure 2-22b

Testing a cable for
leakage

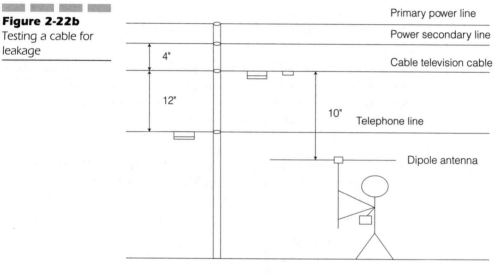

Primary power line

Power secondary line

Cable television cable

4"

12"

10"   Telephone line

Dipole antenna

Table 2-2

Leakage Signal
Level Conversion
to Microvolts
per Meter for
Calculation of CLI

Leak (in dBmV)	Leak (in μV/m)	Leak 2
−39	20	400
−35	30.5	930
−30	54.5	2970
−20	172	29,584
−15	306	93,636
−35	30.5	930
−40	17.2	296
−35	30.6	930
−32	43.2	1866
−30	54.5	2970
		134,512

Summary

The coaxial cable method of television signal distribution consists of a trunk-feeder system where the trunk system acts as a transport for the programming signal to the end of the service area. This trunk system consists of a cascade of trunk repeating amplifier sections based on the unity gain building block. At selected amplifier locations in the populated portions of the service area, bridging amplifiers were installed to provide signals to the feeder cable known as the subscriber distribution portion of the cable system. Taps installed in this distribution cable contain ports for the subscriber drop cable connected to the subscribers' homes. As the feeder cables were extended, distribution amplifiers were installed to increase the signal level needed to connect more subscribers. Several types of amplifying devices are used to allow the cable plant to extend and branch out over the service area. As the frequency of the television carriers increased, the cable loss increased, so amplifier gain at the high frequencies had to increase as well. Low-frequency signals did not need much amplifier gain, which means slope control of the amplifiers is needed. To compensate for this signal-level slope called for a mix of cable equalization networks and amplifier controls. Both methods are integrated in the solid-state integrated circuit automatic gain and slope control amplifiers used in today's coaxial networks.

Two methods of construction are used in coaxial plant serving television signals to subscribers. The first is aerial construction where the coaxial cable is carried on a galvanized steel messenger strand mounted on the utility poles using standard pole-line hardware. It is the cable operator's job to obtain the permits from the pole owners—usually the local power company, the telephone company, or both. To obtain such a permit, the cable operator has to be licensed or have the franchise from the local town, city, or municipality to provide cable television service. The pole owners charge the cable operator for any adjustments in their plant needed to make room on the poles for the cable plant. Once the permits are obtained the cable operator can install such plant on the utility poles; they will be charged an amount of rent per pole per year.

The second method is underground, which means the cable is buried in the ground, placed either directly or in duct or pipe, as the municipality may require. Either a trench has to be opened or a plow method may be used to place the cable system underground. To obtain a permit to install underground plant, all utilities—including power, telephone, water, gas, and sewer—have to have adequate notice so they can locate and mark where their plant is located. Cable electronics and power supplies are mounted in either metal above-the-ground pedestals or in plastic or concrete vaults. Where cable systems operate in cities, the cable operator may have to rent space in municipal ducts and gain access in maintenance holes and or underground vaults.

Maintaining cable plant requires a constant signal-testing program where test ports are placed at critical locations. By using such instruments as signal-level meters or spectrum analyzers, the quality and level of the signal can be measured at these test ports and the data recorded by hand or by computer for later analysis.

At end of construction, a proof of performance test is required by the FCC and should be done to satisfy the cable operator and/or its lending institution that the contractors have built a plant according to the contracted specifications. This test is usually performed before the final payment is made.

The cable connectors are a critical component in coaxial systems. Weatherproofing of connectors keeps moisture problems to a minimum. Heat-shrink tubing is an approved method of water- or weatherproofing the cable connectors, which also gives added mechanical strength to the system. Before installing such heat-shrink tubing the connector must be properly installed to its proper tightness. Loose connectors can allow the signal to escape (signal leakage) and noise and interference to get into the system (signal ingress). Both of these conditions are bad. Since cable operators use frequencies within the cable that are used for normally wireless services such as aircraft communications and navigation used by commercial and military aircraft, any interference caused by cable leakage could be disastrous. The FCC has issued rules and regulations to the cable operator to consistently monitor and control signal leakage, and heavy fines can be imposed for violations. Testing for leakage is a constant and ongoing procedure.

Questions

1. Name the network topology used by a coaxial cable television distribution system.

2. What is meant by a "two-way cable system"?

3. What device is often required for a subscriber to view all cable channels?

4. Name the kind of work the utility companies have to complete before the cable operator can obtain permission to access the utility poles.

5. Cable systems branch out the feeder system into the service area by the use of what devices?

6. Name three main problems that a cascade of amplifiers have that limit how many can be used in cascade.

7. Automatic slope and gain control amplifiers compensate for what parameters in a cascade of amplifiers?

8. What three main types of cable amplifier technology are in use by many cable systems?

9. Explain why the reverse noise buildup is such a large problem in cable systems using a reverse signal band.

10. List the advantages and disadvantages of underground cable construction.

Problems

1. If the output of a bridging amplifier is a constant (flat) signal level of +46 dBmV and feeds a cable section that is 200 ft long, calculate the signal level at the end of the cable section if the loss specifications are 2 dB/100 ft at 750 MHz and 0.5 dB/100 ft at 50 MHz.

2. Given a four-port 26-dB tap device has an input of 44 dBmV, what is the signal level at any one of the four ports?

3. If a 26-dB four-port tap has a through loss of 1.3 dB at 750 MHz and 0.5 dB at 50 MHz, calculate the output level at the through port if the input level is +36 dBmV at 750 MHz and +38 dBmV at 50 MHz.

4. If an 8-dB directional coupler has an input level of +39.5 dBmV at 750 MHz and +33 dBmV at 50 MHz, calculate the signal level at the −8-dB port.

5. If a trunk amplifier has a noise figure of 12 dB and the input signal level is +20 dBmV, calculate the output carrier-to-noise level.

6. At the end of a 20-amplifier cascade of identical amplifiers of the same type in problem 5, what will be the carrier-to-noise ratio in dB?

7. If one trunk amplifier has a carrier-to-composite triple beat (C/CTB) specification of 72 dB, what will be the C/CTB for a 20-amplifier cascade?

8. If it is desired to have a C/CTB specification of 52 dB at the end of an amplifier cascade, how many amplifiers can be in cascade if the single amplifier specification is 63 dB?

9. Calculate the length of one rod of a test dipole antenna that will be used as a cable system leakage test antenna at the carrier frequency of 121.25 MHz.

10. If the dipole antenna in problem 9 is to be used to test for cable leakage and is placed 10 ft from the cable system, what will a signal-level meter read in dBmV at the 20 μV/m?

Headends and Signal Processing

Objectives

After learning the material in this chapter, the student will be able to

- Describe the headend's importance in the cable system network topology.
- Identify problems and solutions of handling the local television stations carried on a cable system.
- Explain the operation of headend electronic signal processing equipment.
- Describe the method and associated equipment for processing microwave and satellite signals.
- Explain the upstream (reverse) systems operation and control at the headend.
- Understand the use of the bidirectional capability of a cable system to control signal security and subscriber billing.

3.1 The Headend

The coaxial cable system headend is the source point for all the programs offered on a system. It is also the main source of control, often referred to as the root of the trunk/branch-tree network. From this point, signals are delivered to subscribers via the trunk or distribution cable system known as the limbs and branches of the tree network topology. The headend system can vary from one central point to a system of other source points feeding a common source node. Headend signal system networks can range from simple to several locations connected by a trunk cable system. Some headend network topology is shown in Figure 3-1.

For cable systems with an active reverse system in operation, reverse signals at the headend have to be processed upon arrival. Processing of reverse signals includes routing of signals to various pieces of equipment used in the cable system business. For example, a subscriber's two-way set-top box may send ordering information to the headend to request selected programming. When such programming has been delivered on the forward or downstream channel, a receipt will be sent back to the headend for subscriber account billing. Upstream services such as meter

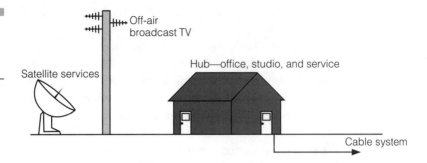

Figure 3-1
Examples of hub/
headend
configurations

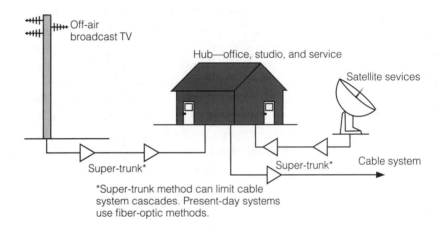

reading, home security system monitoring, and the telephone services supply added income. Cable systems have to offer the local television broadcast stations and often employ several different methods to receive them, including direct off-air reception from a tower-mounted antenna system or an optical fiber connection direct from the station or from a microwave link.

3.11 Local Television Broadcast Stations

Local television broadcast stations provide one very important source of signals for subscribers. These stations are the local network affiliate in the standard VHF band, plus the independents and the public broadcast stations. Local news, weather, and sports are usually favored programming. After all, cable television got its start carrying the off-air broadcast stations to distant areas or areas where the signal quality was poor. Usually cable companies receive off-air broadcast stations from tower-mounted antennas feeding the signal to amplifiers or signal processors in the headend electronic equipment building. This building was often placed near the foot of the tower. An ideal situation would be to have the satellite-receiving antennas and equipment at this location as well.

3.111 Log periodic antennas and Yagi antennas are the two most common types for VHF reception. Figure 3-2 shows the Yagi type of antenna. The log periodic antenna is a bit more broadband and hence can be used for two or three adjacent channels, if the stations are in the same pointing direction. This antenna has good gain, directivity, and high front-to-back ratio characteristics and is usually the antenna of choice for moderate to distant television stations. This type is shown in Figure 3-3. For UHF television station applications, the bow-tie corner reflector can be used for stations not too far away; for distant stations either the wire-type antenna or solid-aluminum parabolic antenna is usually the best choice. These antenna types are shown in Figure 3-4.

Placement on the antenna tower or mast is important and, unfortunately in many instances, is overlooked by some cable systems. The close

Figure 3-2
Yagi pattern

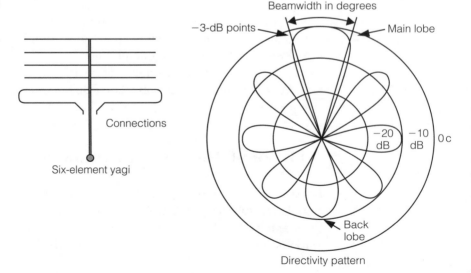

Figure 3-3
Log periodic antenna

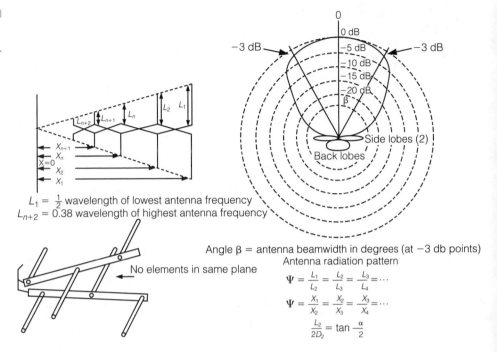

$L_1 = \frac{1}{2}$ wavelength of lowest antenna frequency
$L_{n+2} = 0.38$ wavelength of highest antenna frequency

No elements in same plane

Angle β = antenna beamwidth in degrees (at −3 db points)
Antenna radiation pattern

$$\Psi = \frac{L_1}{L_2} = \frac{L_2}{L_3} = \frac{L_3}{L_4} = \cdots$$

$$\Psi = \frac{X_1}{X_2} = \frac{X_2}{X_3} = \frac{X_3}{X_4} = \cdots$$

$$\frac{L_2}{2D_2} = \tan \frac{\alpha}{2}$$

Figure 3-4
Corner reflector and parabolic antennas

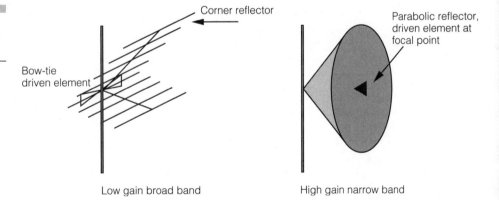

Corner reflector

Bow-tie driven element

Parabolic reflector, driven element at focal point

Low gain broad band

High gain narrow band

proximity of antennas can interfere with the antenna beam patterns of each. Proper spacing, depending on the wavelengths of the antennas, is required to overcome this problem.

3.112 Placement on the tower between VHF antennas is often the main consideration. Since UHF stations operate at shorter wavelengths, the

Figure 3-5
Antenna mounting
location on tower

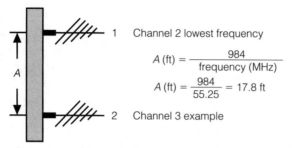

1   Channel 2 lowest frequency

$$A \text{ (ft)} = \frac{984}{\text{frequency (MHz)}}$$

$$A \text{ (ft)} = \frac{984}{55.25} = 17.8 \text{ ft}$$

2   Channel 3 example

1. The antenna for the weakest station should occupy the topmost section of the tower.

2. If the receiver signals are equal, place the highest frequency (smallest) on top so as not to weigh the top unnecessarily.

3. Follow the procedures outlined for proper minimum allowed spacing.

Vertical separation of antennas
(a)

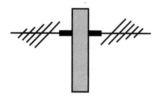

Antennas back-to-back
(180 degrees apart)
(b)

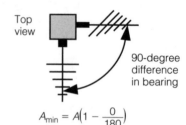

Top
view

90-degree
difference
in bearing

$$A_{min} = A\left(1 - \frac{0}{180}\right)$$

(Antennas 0 degrees apart.)

$$= 17.8 \left(1 - \frac{90}{180}\right)$$

$$= 17.8 \left(1 - \frac{1}{2}\right) = 17.8 \times \frac{1}{2}$$

$$= 8.9 \text{ ft (or half what it was before)}$$

Antennas 90 degrees apart
(c)

separation between antennas can be closer. Also, the location of antennas depends on the pointing direction. For example, antennas placed at directions on opposite tower legs can be spaced closer vertically, as shown in Figure 3-5.

Notice that antennas pointing in a direction 180° from one another can be placed on different tower legs at the same level. When working on the antenna placement problems on a tower, it is key to know the amount of tower height needed for the required antennas. Once this measurement is known, the next issue is the amount of added height needed for proper signal strength. The usual method involves an on-site signal survey, which, of course, involves some expense.

First and foremost is obtaining the broadcast station signal contour maps, which are often available from either the antenna manufacturers or from a consulting firm specializing in performing such on-site signal surveys. Data taken from these maps can indicate whether a full signal survey is necessary. In many instances, the cable system technical staff can simply go to the site and erect a temporary wideband rooftop antenna and take signal-level and picture-quality measurements with a signal-level meter and a TV set. If any interference is evident, further signal tests with a spectrum analyzer may be necessary.

3.113 If some of the desired stations are weak because of the distance from the station, a preamplifier may solve the problem. In some instances, the preamp may be placed in the headend building before the signal processor, or if the signal is extremely weak, the preamp can be placed on the tower at the antenna. Tower-mounted preamps usually are fed power through the coaxial cable down lead. To prevent corrosion problems, care must be taken to maintain good coaxial connections with no chance of moisture ingress. The signal-to-noise ratio is the most important parameter for the preamp placement problem; an example is shown in Figure 3-6.

For UHF stations, the down-lead loss can become a significant parameter for particularly tall towers or where the headend off-air building is separated a long distance from the tower. If this loss is significant, a tower-mounted converter can be used that converts the UHF signal to a VHF signal, which has a lower cable loss. Often, these converters contain a preamp that improves the signal-to-noise ratio. An example comparing the results of the use of a converter and the results without one is shown in Figure 3-7.

Cable systems requiring tower-mounted converters and/or preamplifiers are usually in remote areas, far from the broadcast stations. For metropolitan cable systems, often a tall building's roof can be used to circumvent the multipath reception problem. For such systems, the signal level is not a problem, so small, more elementary antennas are sufficient. Again, an on-site signal survey will tell the story and indicate the needed equipment.

Figure 3-6

Preamp location at headend

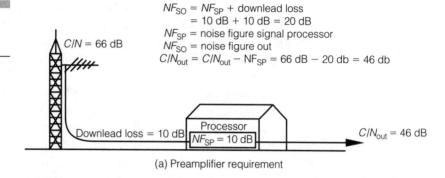

$NF_{SO} = NF_{SP} +$ downlead loss
$= 10$ dB $+ 10$ dB $= 20$ dB
$NF_{SP} =$ noise figure signal processor
$NF_{SO} =$ noise figure out
$C/N_{out} = C/N_{out} - NF_{SP} = 66$ dB $- 20$ db $= 46$ db

$C/N = 66$ dB

Downlead loss = 10 dB

Processor
$NF_{SP} = 10$ dB

$C/N_{out} = 46$ dB

(a) Preamplifier requirement

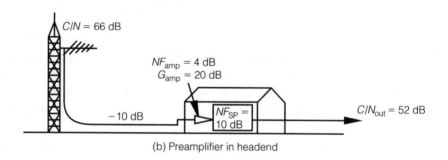

$C/N = 66$ dB

$NF_{amp} = 4$ dB
$G_{amp} = 20$ dB

-10 dB

$NF_{SP} = 10$ dB

$C/N_{out} = 52$ dB

(b) Preamplifier in headend

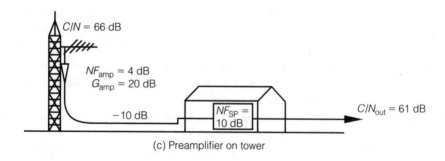

$C/N = 66$ dB

$NF_{amp} = 4$ dB
$G_{amp} = 20$ dB

-10 dB

$NF_{SP} = 10$ dB

$C/N_{out} = 61$ dB

(c) Preamplifier on tower

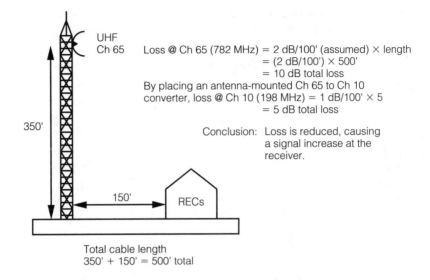

Figure 3-7
Tower-mounted amplifier-converter application and signal-level comparison

UHF Ch 65

350'

150'

RECs

Loss @ Ch 65 (782 MHz) = 2 dB/100' (assumed) × length
= (2 dB/100') × 500'
= 10 dB total loss
By placing an antenna-mounted Ch 65 to Ch 10 converter, loss @ Ch 10 (198 MHz) = 1 dB/100' × 5
= 5 dB total loss

Conclusion: Loss is reduced, causing a signal increase at the receiver.

Total cable length
350' + 150' = 500' total

3.12 Off-Air Reception of Broadcast Stations

Most cable systems use some form of tower or antenna mast to support the off-air antennas. Therefore, some space should be devoted to the discussion of towers and tower maintenance. Towers higher than the usual 200-foot height are considered obstructions and are a hazard to airplane traffic. Proper application should be made to the obstruction department of the regional Federal Aviation Administration (FAA) office. Information about the location of the office can usually be found by looking in the telephone book under "U.S. Government" or by calling to the local regional airport. The FAA office will provide the necessary procedures and application forms for permission to construct. Lighting and painting requirements are specified by the FAA rules. Normally, towers more than 200 feet and less than 300 feet require a flashing red beacon light at the top, a set of three fixed illuminated sidelights at the halfway mark, and orange and white paint alternating in 20-foot intervals. Taller towers may require two sets of fixed sidelights and a flashing beacon at its center, in addition to the top beacon.

Towers fall into two main categories: self-supporting and guyed towers. Self-supporting towers are usually three-legged, but some older ones may be four-legged. They are usually more expensive but use less land than guyed towers. Large self-supporting towers are usually too expensive for most cable television applications. Guyed towers are more common for cable applications and range in height from 100 feet to 300 feet. Higher towers require more land for the guy wires and often use three inner guys and three outer guys. A comparison between the height and land requirements of a freestanding tower and a guyed tower is shown in Figure 3-8.

Guyed towers are usually triangular and may have either a fixed-based mount or a tapered-based mount, as shown in Figure 3-9. The difference

Figure 3-8
Headend towers

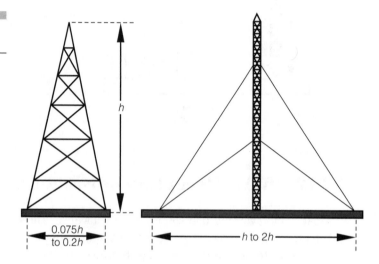

Figure 3-9
Tower-base mounting

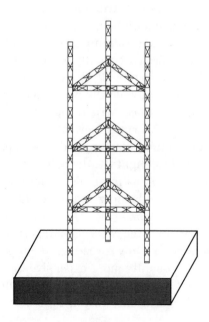
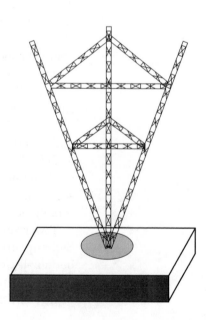

is in the distribution of the bending moments or forces on the tower legs. Shorter guyed triangular towers often use the fixed-base tower. Solid-rod towers, which are usually a lot heavier, use the tapered mount. Tower manufacturers typically have a customer engineering service department available to assist the customer in specifying an appropriate tower. Tower installation crews can often be arranged through the manufacturers. In many instances, a tower turnkey contract can be obtained that provides engineering, installation, grounding, lighting, and antenna installation services. Such a contract is often the best way for a cable system operator because most cable systems do not have a staff member trained in tower work.

3.121 Proper tower grounding is an important consideration in the control of thunderstorm lightning. After all, the tower is a big lightning rod itself. Some operators just ground the base of the tower to a system of ground rods. A base grounding method is shown in Figure 3-10. In areas prone to heavy lightning, a sharp copper rod is mounted at the top of the tower and is connected through heavy copper wire to the ground point at the base of the tower. The four ends are tied together by placing heavy copper strips in the concrete before the tower base is poured. Placing the

Figure 3-10
A typical tower-grounding method

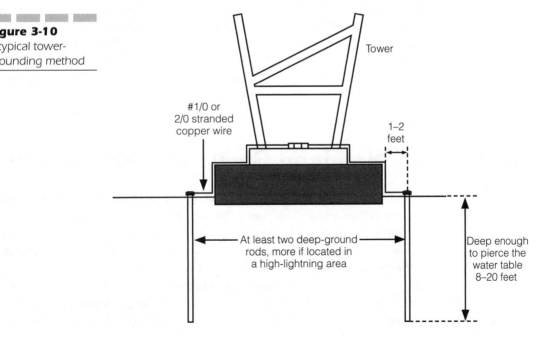

tower base on top usually provides a superior ground for the tower. Testing the ground with an instrument known as a megger will indicate the effectiveness of the ground. If this is performed at the time of installation, subsequent measurements on a yearly basis will detect any deterioration of the tower ground system. If a tower requires lighting, a commercial electric service has to be connected all the way up the tower. Thus, a good and effective ground is necessary to protect the lighting system and any tower-mounted electronic devices such as preamplifiers and converters.

3.122 Tower lighting systems are usually available through the tower manufacturer. The flashing control system is usually mounted near the tower base on one leg of the tower and has a light-detecting cell that turns the light system on at dusk. Because the tower lights use considerable electricity, it is prudent to just turn them on at dusk and off at sunup. The flashing mechanism is also mounted in the control box. Early flashing systems were electromechanical; a motor turned a crank that opened and closed the beacon switch. Most present-day systems use solid-state flashing technology.

Periodic inspection of the tower system should be performed on a seasonal (four times per year) basis. Such an inspection should include the guying system, as well as the anchors and wire fastenings and grounding. Any loose clasps should be tightened and any observable rust should be coated with protecting spray. The lighting system can be easily checked by placing a heavy work glove over the daylight-sensing element, thus turning on the beacons and the sidelights. Burned-out bulbs should be promptly replaced. Also, the tower grounding system should be inspected and any observable corrosion treated. A record of these inspections should be kept in a tower maintenance file.

3.2 Headend Electronic Equipment

Signal-processing systems for received off-air television signals have changed drastically throughout the years—and for the better. Early systems simply used single-channel amplifiers operating with vacuum tubes as the gain element. With the development of transistors that exhibited high-frequency gain characteristics, the single-channel strip amplifiers greatly improved. The addition of automatic gain control was also a welcome addition. A major improvement in off-air signal processing methods occurred with the development of the heterodyne signal processor. Signal processors today use integrated circuit (IC) technology and therefore are physically smaller, consume less power, and operate superbly.

3.21 Strip Amplifier Equipment

The strip single-channel amplifier with automatic gain is probably the lowest-cost method to use for off-air signal processing. Essentially, this method couples a gain block with channel band-pass filters. At this point, the only signal amplified is the channel passed by the band-pass filter. The pass band is usually flat throughout the 6-MHz-wide television channel. The signal level is adjusted by a front panel control. Both the video and audio signals are passed through the amplifier, but an audio 4.5-MHz trap allows the sound carrier to be set to the usual 15 dB below video carrier level. This is necessary for consecutive channel operation and prevents the sound carrier level from interfering with the upper adjacent picture carrier.

Strip amplifiers often appeared in early systems and were often replaced later with the heterodyne signal processor. The strip-amplifier method is shown in Figure 3-11.

3.22 Heterodyne Signal Processing

Most systems use heterodyne signal processors for off-air television broadcast services. At one time, some cable systems opted for the demod/remod scheme, which required the received television station to be demodulated to base band and remodulated back to either the same or a different cable channel. Such a system allowed local messages and or advertisements to be inserted at the video signal level. Also, an emergency alert system (EAS) could be inserted in all the television channels. Voice or audio signal override could be easily introduced. Today, many manufacturers of heterodyne signal processors offer features similar to the demod/remod

Figure 3-11
Strip amplifier

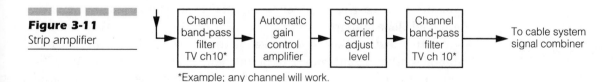

*Example; any channel will work.

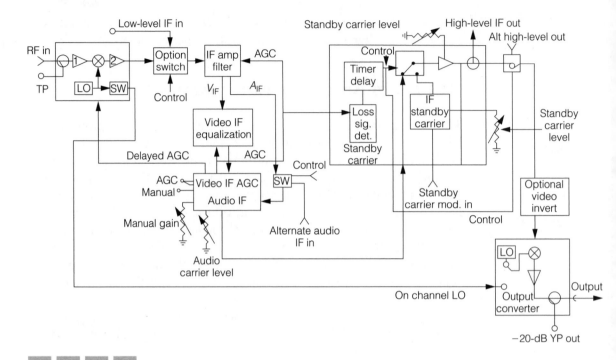

Figure 3-12
Block diagram of a signal processor

method, but with signal switching at IF. A block diagram of a heterodyne signal processor is shown in Figure 3-12.

This type of processor has many desirable features that enable several options for the operator. Probably the most desirable feature is the standby carrier. If the input signal is lost, either because the station went off the air or a problem occurred with the receiving antenna, a substitute carrier at the correct level would be introduced into the cable system. This replacement signal would continue until the off-air signal was returned to the input of the processor. This standby carrier could be modulated by a replacement video signal, such as a simple "please stand by" message. A properly level-adjusted standby carrier is absolutely required for the system's pilot carrier in order to maintain the cable system automatic level and slope system.

Another feature that is often used is the alternate IF switching, which allows another signal source to be introduced at the IF frequency. This feature was developed for cable systems that were required to drop an

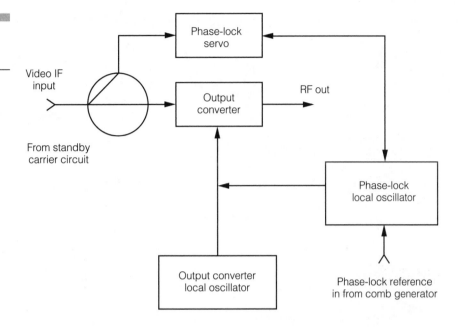

Figure 3-13
Signal processor
phase-lock unit

off-air channel that contained the same program as one of the must-carry broadcast stations. Such a condition is referred to as non-duplication, or simply "non-dupe," switching. Cable systems that use the off-air method of receiving broadcast programs use heterodyne signal processors. Several manufacturers make some superb signal-processing equipment that also features the capability of phase locking to a comb generator. Such phase locking is necessary for either incrementally related carrier (IRC) or harmonically related carrier (HRC) controlled systems. A block diagram of the processor phase lock loop is shown in Figure 3-13.

3.23 Television Signal Modulators

A television modulator, as used by the CATV operator, is in essence a miniature television broadcast transmitter. The signal-processing stages have to perform with the same specifications as a broadcast television station. The block diagram for a present-day television modulator is shown in Figure 3-14. Elementary television modulators are found in such devices as VCRs, where the recorded video and audio signal is played and sent to a modulation circuit that transforms it to a standard television channel,

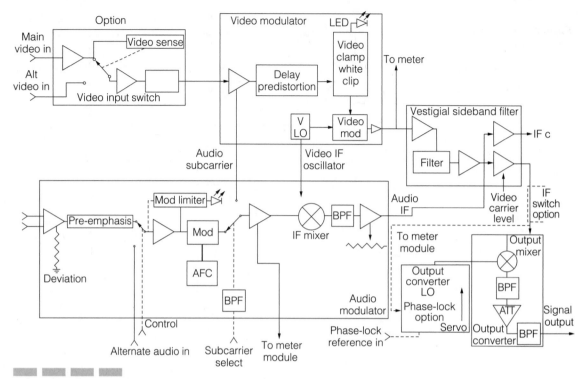

Figure 3-14
Television modulator

typically channel 3 or 4. The modulator for cable television use is vastly different from this type of device.

In Figure 3-14, notice that there are two video inputs and two audio inputs. One is the main channel and the other is the alternate channel, which is controlled by a selector switch. The switch can also have the option to switch to the alternate source upon loss of the main input video signal. The alternate source can be a simple "please stand by" or some other appropriate message. Just about any signal can either be introduced or outputted for a variety of applications. Options provide for full video/audio metering and system phase-locking capability.

Manufacturers of headend equipment offer cable systems a variety of television modulators from extremely compact types to larger ones containing a large selection of options. Some modulators have a selectable or tunable output and can be considered as a sort of universal device. Often these modulators do not have a selectable band-pass output filter and thus can add broadband noise into a channel-combining network. A more prudent approach is to use normal single-channel output converters with band-pass filtering for all channels that require a modulator. The tunable-type modulators can be used as spares. In general, modulators are used for satellite-originated and locally generated program channels.

3.24 Signal-Combining Methods

The next step in assembling the television channel lineup is the combining of all the program channels into one terminal used to drive the coaxial cable network. Care must be taken at this point to not inject any added noise or unwanted signals into the system. In some instances, through either local ordinances or for economic reasons, the cable television system may use an active AM or FM tower for its off-air antenna system. Thus, the problems of preventing the transmitted station from entering the cable system can be extremely difficult. An example of such a problem is shown in Figure 3-15, but the solution may not be as simple as that shown. Situations such as this should be avoided if at all possible. A spectrum analyzer measurement of the whole cable spectrum at the combining point will show all the video and audio carrier levels as well as any extraneous signals.

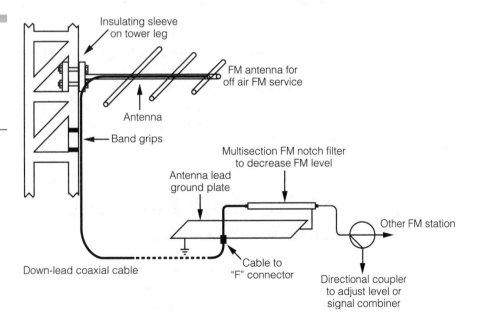

Figure 3-15
FM antenna for broadband FM service on cables. The antennas are mounted on an FM-transmitting tower.

Figure 3-16

Signal-combining
network types

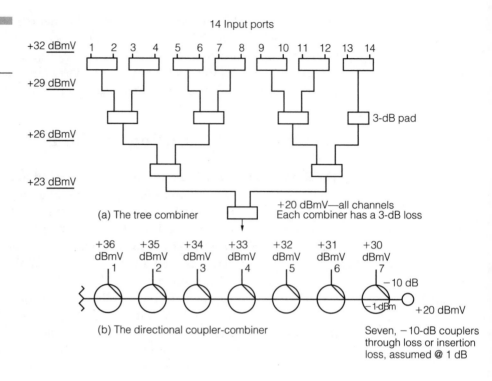

(a) The tree combiner

+20 dBmV—all channels
Each combiner has a 3-dB loss

(b) The directional coupler-combiner

Seven, −10-dB couplers
through loss or insertion
loss, assumed @ 1 dB

The combining network can be made from a tree of 75-ohm two-port splitters that is used as a combiner or as a series of directional couplers. The tree method contains the least loss and therefore the signals to be combined do not have to run at such high levels. Manufacturers of headend equipment usually have directional couplers, which are most frequently used for combining networks in their product lines. Ideally, each channel should have an output band-pass filter. This prevents any extraneous broadband noise from entering the cable system. These two types of combining methods are shown in Figure 3-16.

3.3 Headend Powering and Monitoring

Cable television headends usually are operated by commercial power brought to the site by aerial or underground means. In the case where there is a commercial power failure, some means of backup power is recommended. Current cable systems should attempt to operate with the same reliability figures as the telephone system. This is a must-do if cable operators are going to compete in the business of providing voice, data, and video communications. Since signal reliability is an issue, headend signal monitoring also should be considered.

3.31 Headend Emergency Powering Methods

Standby powering of headend equipment falls into two basic categories: an engine-generator set and a battery backup set. The battery backup set can be as exotic as an uninterruptible power supply (UPS). Since most headends have some kind of alarm system, such as for unauthorized intrusion, fire, or smoke, these could also alert the loss of commercial power. Service personnel could then go to the headend site to assess the situation. The most common type of headend emergency power is the engine generator set, simply because it can usually provide 100 percent of the usual power load, which includes air-conditioning and tower lights in addition to the normal electronic load. The engines are usually reciprocating multicylinder engines that operate on a variety of fuels. Large standby plants often use diesel fuel, while others use gasoline, propane, or piped-in natural gas.

Several manufacturers of engine-driven standby power plants offer equipment in a broad range of powers. Battery-powered backup systems usually provide enough power to cover anywhere from a one- to two-hour off period. For engine-driven emergency power systems, an automatic program changeover switch is the method of choice. This device starts the engine, tests the voltage, and transfers the power. Upon return of commercial power, as sensed by the switch, the power transfer is made back to commercial power and the engine is shut down.

An engine-run hour meter records the amount of time the engine was in operation. This data should be recorded in the plant maintenance file. Some emergency electric plant controllers have a weekly or monthly cycle period where the engine starts and runs for a predetermined period and then shuts down. This so called exercise cycle may or may not actually transfer the electrical load.

Battery backup power systems usually do not have an exercise cycle. The battery pack's condition should be checked visually as well as electrically, and the condition of the charge should be recorded in the system maintenance files. Most uninterruptible power backup systems either remain online, providing power while simultaneously being charged, or are switched online at commercial power outages. At the return of commercial power service, these systems transfer off-line and resume charging. As yet, most manufacturers of headend electronic devices do not offer backup power options on a device-by-device basis. Since many computer systems use such backup power supplies, it is most likely this method will be used in hubs and headends.

3.32 Signal Monitoring and Testing

Signal reliability is now of extreme importance in many modern systems, so some means of signal monitoring and testing should be used. A test technician, who records the data in a site log, can do the testing and monitoring program manually. The advantage of such a program is that if a problem is observed, it can be corrected at once. However, this labor-intensive method is the most expensive. An automatic method involves the installation of programmed instruments to make periodic measurements in a predetermined sequence. The results can be converted to digital data and recorded on a floppy disc. This system can also be interrupted manually at the control computer terminal for a visual test of the systems operating parameters.

If one or more headend sites are monitored in this fashion, a central office monitoring computer can be operated via a data-communication channel through the cable system. Such an interconnected system is shown in Figure 3-17. Each site can in turn be interrogated for its measurement data by the control computer through the usual data-type modems. Data found to be faulty by the monitoring computer will signal an alarm so technical personnel can respond and correct the problem. Carefully selecting the alarm threshold can facilitate heading off problems before an actual signal outage occurs.

3.33 Headend Grounding and Bonding Considerations

Proper grounding and bonding are important concerns for headend electronic equipment. For most headend layouts, electronic devices such as signal processors, modulators, satellite receivers, ad-insertion equipment, and so on are mounted in metal racks. In some instances, open standard, 19-inch telephone-type racks are used; for more upscale head-

Figure 3-17

Status monitoring
of end levels and/or
power-supply status

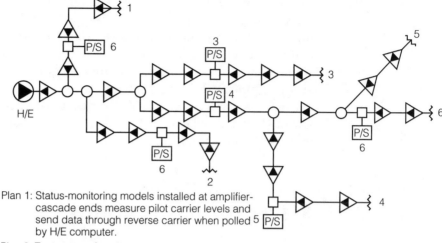

Plan 1: Status-monitoring models installed at amplifier-
cascade ends measure pilot carrier levels and
send data through reverse carrier when polled
by H/E computer.

Plan 2: Test status of each power supply when polled
by H/E computer.

Plans 1 and 2 could both be used.

ends the equipment is mounted in enclosed rack cabinets. The metal racks should be connected together with heavy (1/0 or 2/0) copper wire, bus bar, or braids, and then connected to a grounded electrode. This ground point should be the same electrode as the power system ground. Now, no-fault currents can be set up through several paths to ground. The ground method should provide less than 25 ohms to ground, as tested with megger or milliohmmeter.

Some headends operating near television or FM/AM radio-transmitting facilities have to use RF chokes and bypass capacitors at the power service panel to prevent RF ingress of unwanted signals entering the cable system. Cable entering the electronic equipment building from the antenna system, referred to as down leads, have to be connected to a heavy aluminum plate grounded at the headend common ground point. The heavy, usually 1/2-inch solid aluminum-sheath system cable used as down leads is connectorized to this panel through half-inch "F" connectors. Now, simple RG-6 cable connects to the equipment in the racks.

Since most headends use towers for the off-air receiving antennas, lightning can again be a problem. Proper grounding can help prevent the direct effects of a lightning strike, but large spikes of voltage can be introduced into the power lines, causing power surges of significant magnitude. Such power surges can cause blown fuses or equipment failure of a significant nature. A good idea is to have a surge-suppression system in place. Simple surge suppressors can be used at each device plug, but a larger unit mounted in or at the power distribution panel saves a lot of clutter and is quite effective. Several manufacturers make panel-mounted surge suppressors offering various features, shapes, and sizes. Basically, the surge suppressor has to pass the load, removing any voltage spikes and delivering smooth power free of electrical transients to the equipment. By now it should be evident that there are a large number of considerations when planning a headend.

3.4 Satellite Systems

Some of the people who started cable systems back in the 1950s to 1960s recall the days when there were no operating satellite systems. Cable television systems were simply a means used to deliver greatly improved television broadcast signals to subscribers, a true community antenna television system. Many early systems had unused bandwidth or channel space. When satellite systems were developed to the point where more programming became available to the cable operator, more bandwidth was needed to carry the added services. The axiom "If you have more to sell, you sell more" became an important rule for cable operators. Communication satellites are very important to the telecommunication industry and affect many users around the world.

3.41 Satellite System Receiving Considerations

Of the many satellite systems in place covering the western hemisphere, the geostationary ones are most important to the cable television industry. A geostationary satellite has a circular, equatorial orbit, as shown in Figure 3-18. Notice that the usual orbit direction is the same as the Earth's rotation and the height above the Earth's surface is about 36,000 kilometers (21,600 statute miles). One such satellite can cover more than two-thirds of the Earth's surface. Satellites that orbit the Earth have to conform to the velocity equation given in Figure 3-18. Satellites in a geostationary orbit have the same angular velocity as the Earth. Therefore,

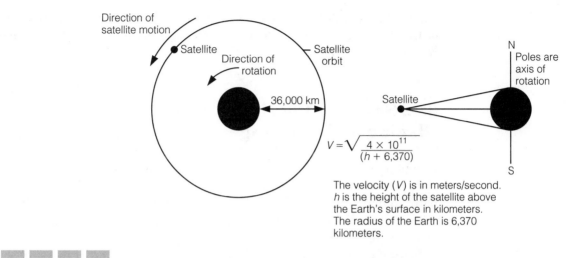

Figure 3-18
Geostationary satellite orbit

the circumference of the orbit divided into the orbit speed as calculated by the equation will give the period of the orbit as 24 hours. This is the same as the Earth's period of rotation.

When planning a satellite-receiving system for a cable system, a frequency-coordination study should be made as soon as the desired satellite channels have been determined. This study provides information on the amount of expected or possible interference from any other surrounding microwave-communication facilities. Microwave-common carriers, such as any of the local telephone systems, use the same frequency bands as some of the satellite systems. The frequency-coordination study is a service offered by several companies, which have amassed a database of frequencies and their propagation paths. If this study indicates possible interference from any local sources, an on-site measurement may be necessary to determine if the interference is of a sufficient level to cause a problem. A possible solution is to erect a conductive shield near the receive antenna or to construct an artificial hill in the direction of interference. Some manufacturers make a shield ring around the parabolic antenna, which reduces the effective side lobe pattern of the antenna. If no solutions are feasible, the receiving site may have to be relocated to a place that has no interference problems. Now it may be necessary to super-trunk the signals to the hub/headend site or main signal distribution location. All of this investigative work should be satisfactorily completed before any construction begins.

3.42 Satellite Receiving Antenna Placing and Pointing

One important task is to make sure the parabolic receiving antenna can get a clear path to the satellite. No trees or buildings should obstruct the path. In many instances, trees have to be removed to clear the path from antenna to satellite. In order to do this, the azimuth and elevation look angles have to be determined.

Several pieces of information have to be assembled in order to establish the look angles for the antenna. The satellite longitude has to be known, but there is no latitude given because zero latitude is the equator. Next, the receiving-site latitude and longitude have to be determined to an accuracy of the nearest hundredth of a degree. A U.S. coast and geodetic survey map of the 7.5-minute type is adequate to determine the site's latitude and longitude coordinates. Often, sporting goods stores have these maps for sale. A local civil engineering or surveying office will be able to determine the site coordinates, or a handheld GPS satellite global-positioning receiver operating in the differential mode can provide coordinate information to a sufficient accuracy. Such instruments are available in sporting goods stores for a few hundred dollars. In most cases, the prospective site coordinates are needed for the frequency coordination study. The calculation for the azimuth angle in degrees from true north and the elevation angle in degrees from level ground can be calculated. A completed example of the method and formulas is shown in Example 3-1. Once the elevation angle is found, the slant range, or simply the distance from the receiving antenna site to the satellite, can be determined, as shown in Example 3-2. This distance is used to calculate the signal path loss.

For various satellites, the satellite communication operators have a map with the projected effective isotropic radiated power (EIRP) contour lines appearing on it. For receiving sites found on the map, the EIRP can be determined. An example of such a map is shown in Figure 3-19. The calculation for the expected carrier-to-noise level at the receiver preamplifier input can now be made, and a worked-out example is shown in Example 3-3. If a better C/N ratio is required, a larger antenna will be necessary.

Example 3-1

The formulas are as follows:

Antenna azimuth angle (degrees)

$$= 180 + \tan^{-1}\frac{(\tan D)}{(\sin X)} \text{ (from true north)}$$

Antenna elevation angle (degrees)

$$= \left(\frac{\{\cos D \times \cos X\} - 0.15126}{\sqrt{\{\sin D\}^2 + \{\cos D \times \sin X\}^2}}\right)$$

where X = Earth station site latitude
Y = Earth station site longitude
Z = satellite longitude
$D = Z - Y$

Example

Earth station site latitude 41°38′11.0″ N
Earth station site longitude 70°27′42.0″ W
Satellite Galaxy I longitude 93.5° W
Solution: Convert latitude and longitude of Earth station site to decimal degrees.

Earth Station Latitude		Earth Station Longitude	
41° =	41.0000	70° =	70.0000
38′ = 38/60=	0.6333	27′ = 27/60 =	0.4500
11.0″=11/3600 =	0.0031	42″ = 42/ 3600 =	0.0117
	41.6364 = 41.64		70.4617 = 70.46

where X = 41.64°
Y = 70.46°
Z = 93.5°
D = 93.5 − 70.46 = 23.04°

$$\text{Azimuth angle (degrees)} = 180 + \tan^{-1}\left(\frac{\tan D}{\sin X}\right)$$

$$= 180 + \tan^{-1}\left(\frac{\tan 23.04}{\sin 41.64}\right)$$

$$= 180 + \tan^{-1}\left(\frac{0.425}{0.66}\right)$$

$$= 180 + \tan^{-1}(0.644)$$

$$180 + 32.8 = 212° \text{ from true north}$$

$$\text{Elevation angle (degrees)} = \tan^{-1}\left(\frac{\{\cos D \times \cos X\} - 0.15126}{\sqrt{\{\sin D\}^2 + \{\cos D \times \sin X\}^2}}\right)$$

$$= \tan^{-1}\left(\frac{\{\cos 23.04 \times \cos 41.64\} - 0.15126}{\sqrt{\{\sin 23.04\}^2 + \{\cos D \times \sin X\}^2}}\right)$$

$$= \tan^{-1}\left(\frac{\{0.920 \times 0.747\} - 0.15126}{\sqrt{\{0.391\}^2 + \{0.920 \times 0.664\}^2}}\right)$$

$$= \tan^{-1}\left(\frac{0.6872 - 0.15126}{\sqrt{0.153 + 0.373}}\right)$$

$$= \tan^{-1}\frac{0.5359}{0.723}$$

$$= \tan^{-1}(0.7389)$$

$$= 36.5°$$

Example 3-2

The distance from a point on Earth to a satellite can be calculated using the following formula:

$$d = \sqrt{[(r + h)^2 - (r(\cos \theta))^2]} - r(\sin \theta)$$

where d = distance to the satellite in kilometers
r = radius of the Earth = 6370 kilometers
h = height above satellite above equator = 35,780 km
Θ = elevation angle in degrees

Example: $\Theta = 36.5°$

$$d = \sqrt{(42.15 \times 10^3)^2 - ((6.37 \times 10^3)(\cos 36.5))^2} - (6.37 \times 10^3)(\sin 36.5)$$

$$= \sqrt{1806 \times 10^6 - (5.12 \times 10^3)^2} - 3.79 \times 10^3$$

$$= \sqrt{1806 \times 10^6 - 26.2 \times 10^6} - 3.79 \times 10^3$$

$$= \sqrt{1780 \times 10^6} - 3.79 \times 10^3$$

$$= 42.2 \times 10^3 - 3.79 \times 10^3$$

$$= 38.4 \times 103 \text{ km}$$

$$= 38,400 \text{ km}$$

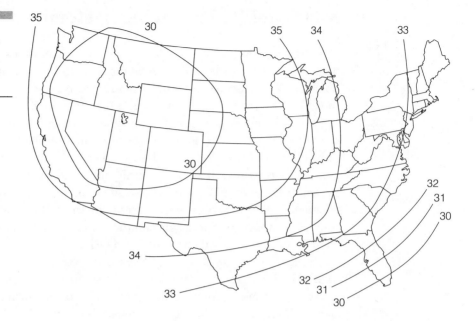

Figure 3-19
Calculation of the distance from the receiving antenna site to the satellite

Example 3-3

The path between a satellite and a receiving site on Earth may be calculated using the following procedure:

$$L_{\text{path}}(\text{dB}) = 32.44 + 20 \log d_{(\text{km})} + 20 \log f_{(\text{MHz})}$$

Example: For a frequency of 4 GHz (4000 MHz) and where
$d = 38{,}400$ km.

$$L_{\text{path}}(\text{dB}) = 32.44 + 20 \log 38{,}400 + 20 \log 4000$$

$$= 32.44 + 20 \log 3.84 + 80 \text{ dB} + 20 \log 4 + 60 \text{ dB}$$

$$= 32.44 + 11.69 + 80 \text{ dB} + 12 + 60$$

Path loss $= 196$ dB

3.43 Satellite Receiving Systems

Once the calculations are made, the receiving-antenna specifications will be known. Now the decision can be made as to the antenna size, the low-noise block down-converter (LNBC), and the type of feed-mount and lead-in cable. Several types of antennas can be fitted with feed mounts covering several satellites depending on their spacing in the satellite arc. For certain applications, 5- and 6-meter antennas can be fitted to cover three different satellites that are closely spaced. Because adjacent satellite channels are transmitted alternately polarized (odd-numbered channels horizontally polarized, even-numbered channels vertically polarized), the frequency difference between adjacent channels and this polarization difference prevents any adjacent channel interference. Table 3-1 lists some commonly used C-band channel information.

Satellite-receiving antennas have been fitted with the required vertically and horizontally polarized feeds with a low-noise block down-converter connected to each one. The converter performs two basic functions. The first is to amplify the weak microwave signal, and the second is to convert the signals to a lower frequency. Converting to a lower frequency results in lower cable loss and therefore a less attenuated signal at the receiver. This method is shown in Figure 3-20 and is the most commonly used method.

3.44 Satellite Channel Converting Equipment

The receiver takes the converted C-band signals and produces a baseband descrambled signal. Many manufacturers provide what is known in the industry as an IRD (integrated receiver descrambler) satellite

Figure 3-20

Satellite antenna block converter configuration

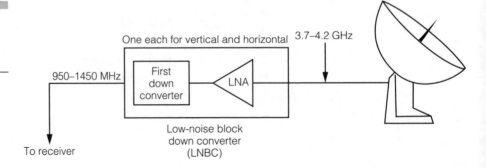

One each for vertical and horizontal 3.7–4.2 GHz

950–1450 MHz

First down converter

LNA

Low-noise block down converter (LNBC)

To receiver

Table 3-1

C-Band Satellite
Channel
Information

Satellite Channel	Downlink (in MHz)	Downlink Polarization	
		A[a]	B[b]
1	3720	V	H
2	3740	H	V
3	3760	V	H
4	3780	H	V
5	3800	V	H
6	3820	H	V
7	3840	V	H
8	3860	H	V
9	3880	V	H
10	3900	H	V
11	3920	V	H
12	3940	H	V
13	3960	V	H
14	3980	H	V
15	4000	V	H
16	4020	H	V
17	4040	V	H
18	4060	H	V
19	4080	V	H
20	4100	H	V
21	4120	V	H
22	4140	H V	V
23	4160	V H	H
24	4180	H V	V

[a]Satellites with polarization listed in Column A are GE Satcom C_3 and C_4 and Hughes Galaxy IX.

[b]Satellites with polarization listed in Column B are Hughes Galaxy IV, V, VI, VII, I-R, G.E. Satcom C_1, and G.E. American GE-1.

Table 3-2

C-Band to L-Band Conversion

Transponder Number	C-Band Frequency (in MHz)	L-Band Frequency (in MHz)	Transponder Number	C-Band Frequency (in MHz)	L-Band Frequency (in MHz)
1	3720	1430	13	3960	1190
2	3740	1410	14	3980	1170
3	3760	1390	15	4000	1150
4	3780	1370	16	4020	1130
5	3800	1350	17	4040	1110
6	3820	1330	18	4060	1090
7	3840	1310	19	4080	1070
8	3860	1290	20	4100	1050
9	3880	1270	21	4120	1030
10	3900	1250	22	4140	1010
11	3920	1230	23	4160	990
12	3940	1210	24	4180	970

receiver. The receiver descrambler circuits are integrated into one package, thus conserving rack space in often crowded electronic equipment bays in larger headends. Table 3-2 lists C-band channel frequencies to their down-converted values. Notice that the low-frequency C-band signals are converted to high-band frequency signals. This is due to the heterodyne frequency conversion process. The L-band receivers on the market today have the C-band channel number indicated on the front panel, making the frequency values superfluous. An L-band receiver is shown in Figure 3-21.

3.45 Satellite Television Scrambling

Much of the programming on satellite channels is scrambled and/or encrypted using several methods. General Instrument Corporation (GI) developed the early type called Videocipher II, which was upgraded to

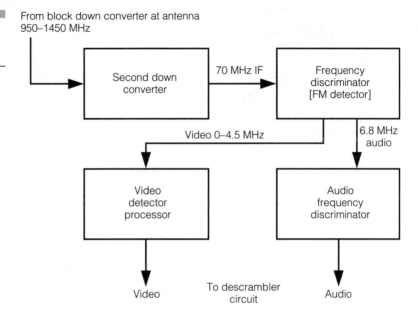

Figure 3-21

L-band to IF conver-
sion in receiver

Videocipher II+. Later methods that were developed and used by GI are DCI, DCII, DCII MCPC, and DCII SCPC. Scientific Atlanta Corporation has a series that is available to program suppliers called Pwrvu, Pwrvu MCPC, and Pwrvu SCPC. The purpose of using any of these methods of encoding is to prevent the unauthorized use of programming material. The key for decoding is contained in the authorized equipment in the cable system headend. The cable operator can then re-scramble in the format compatible with the systems' subscriber converter box.

One might ask, if the signal is already scrambled, why de-scramble and then re-scramble? The answer depends on who is doing the scrambling or protecting the program service. Since passing the signal flow through more circuits adds noise or distortion, possibly a better arrangement would be to pass the scrambled signal through the cable system to the subscriber's set-top box, where the signal is programmed to decode or unscramble. Doing it this way, the program supplier knows the cable company that is providing the service to the subscriber.

Grounding the Earth station antennas is also important in order to protect the sensitive LNBC equipment from the effects of lightning. Such a grounding plan is shown in Figure 3-22.

Figure 3-22
Satellite antenna−
mounting platform
and grounding
system

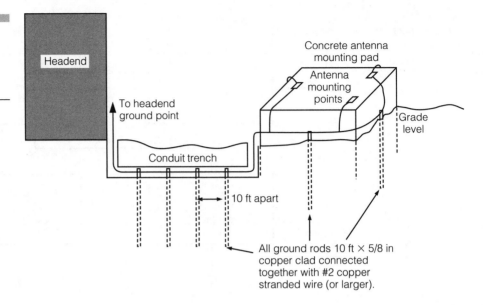

3.46 Insertion of Satellite TV Channels on the Cable System

At the master headend site, the satellite-received signals are either sent clear as part of a basic program package or are re-scrambled as a premium or pay channel. All satellite channels are modulated on a cable system channel and combined with the off-air broadcast stations, plus any of the locally generated channels. In general, locally generated programs are the local television studio−originated signals, a character generator with the listing of the programs (program guide), and a character generator with lost-and-found or news announcements.

A block diagram showing the device interconnections for the processing of satellite program signals is shown in Figure 3-23. Many options and choices are available within the methods of interconnecting the equipment required to assemble the program lineup. Whatever the choice, the signal quality must always be preserved simply because it can never be better at the headend. Once the signal enters the cable system it is naturally degraded and much now depends on the cable system design.

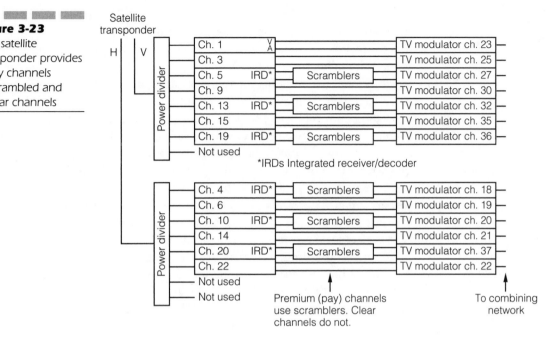

Figure 3-23

One satellite transponder provides 6 pay channels re-scrambled and 7 clear channels

3.5 Microwave Systems Applications

Many cable system operators often find that in some instances a microwave link is either the only method or the least expensive method to import or distribute programming signals from hub to hub, such as when there are no roads or rights of way to feed the hub with an input signal. A cable system wishing to distribute its signal to an area separated by a deep gorge should find a microwave hop to be ideal. Either the whole downstream service band can be up-converted or groups of programming signals can be converted to microwave frequency bands and transmitted across the gorge to another service area. Because microwave links are often bidirectional, the reverse signals from the remote area can be accommodated. The cost factors of such microwave links can vary from inexpensive for one bidirectional short hop to very expensive for multiple hops of long distances.

The success of a microwave system depends on careful planning and design of the link. This work consists of analyzing the link terrain, path access, tower permits, and equipment selection. Some large cable multiple service operators (MSOs) have the technical staff to do the job. There are microwave contracting companies available who will often perform a turn-key operation. To appreciate the complexity of such a project an example of planning a microwave link is given in this section. Table 3-3 lists microwave frequency bands for a cable operator's use. Cable systems that are considering a microwave connection have to go through an FCC licensing procedure that requires filing an application. It is often more expeditious to engage the services of a law firm that specializes in working with the FCC. In many instances, such law firms can refer cable operators to high-quality contractors. The FCC licensing procedure is similar to that for a satellite-receiving station in that a frequency-coordination study has to be performed. This study analyzes the existing microwave paths and their operating frequencies to investigate any interference problems. The result of this study is a required portion of the FCC application process. All permits—including financial, FCC license, and municipal permits—have to be in place before actual construction is allowed. At the conclusion of construction, a turn-on and testing phase followed by a proof-of-performance test has to be completed before normal operation can commence.

3.51 Microwave Master Headend Signals to Hub Sites

This may consist of the usual cable television headend located on a prime site and transmitting these signals to other hub sites (other CATV systems) located miles away in several directions via a microwave line of site (LOS) system of the amplitude-modulated link (AML) type. Also, some cable television systems may receive one or more channels or programming via a common-carrier microwave-service company. Usually when a cable television operator contracts to receive programming via a common-carrier microwave company, this company, as part of the contract, provides the working link, both equipment installation and maintenance, for a fixed monthly fee. Usually all the cable operator has to provide is headend building space, tower space, and electric power. Common-carrier microwave companies often choose FM technology. The systems cable operators have to maintain and service their own AML-type microwave systems. The AML system takes the cable television band of carrier frequencies and converts this band to microwave frequencies and transmits the signal to the receiving site, where the microwave band of frequencies is converted back down to the standard cable television band. A listing of the AML conversion frequencies is shown in Table 3-3. There are four

Table 3-3

CARS Band
Frequencies

AML Frequencies (Bandwidth – 6 MHz)

CATV Channel	VHF Boundaries (in MHz)	CARS Group C		CARS Group D		CARS Group E		CARS Group F	
		FCC Designation	Microwave Boundaries (in MHz)	FCC Designation	Microwave Boundaries (in MHz)	FCC Designation	Microwave Boundaries (in MHz)	FCC Designation	Microwave Boundaries (in MHz)
2	54–60	C01	12700.5–12706.5	D01	12759.7–12765.7	E01	12952.5–12958.5	F01	13012.5–13018.5
3	60–66	C02	12706.5–12712.5	D02	12765.7–12771.7	E02	12958.5–12964.5	F02	13018.5–13024.5
4	66–72	C03	12712.5–12718.5	D03	12771.7–12777.7	E03	12964.5–12970.5	F03	13024.5–13030.5
PT	72–76	C04	12718.5–12722.5	D04	12777.7–12781.7	E04	12970.5–12974.5	F04	13030.5–13034.5
5	76–82	C05	12722.5–12728.5	D05	12781.7–12787.7	E05	12974.5–12980.5	F05	13034.5–13040.5
6	82–88	C06	12728.5–12734.5	D06	12787.7–12793.7	E06	12980.5–12986.5	F06	13040.5–13046.5
FM1	88–94	C07	12734.5–12740.5	D07	12793.7–12799.7	E07	12986.5–12992.5	F07	13046.5–13052.5
FM2	94–100	C08	12740.5–12746.5	D08	12799.7–12805.7	E08	12992.5–12998.5	F08	13052.5–13058.5
FM3	100–106	C09	12746.5–12752.2	D09	12805.7–12811.7	E09	12998.5–13004.5	F09	13058.5–13064.5
FM4	106–108	C10	12752.5–12754.5	D10	12811.7–12813.7	E10	13004.5–13006.5	F10	13064.5–13066.5
AUX1	108–114	C11	12754.5–12760.5	D11	12813.7–12819.7	E11	13006.5–13012.5	F11	13066.5–13072.5
AUX2	114–120	C12	12760.5–12766.5	D12	12819.7–12825.7	E12	13012.5–13018.5	F12	13072.5–13078.5
A	120–126	C13	12766.5–12772.5	D13	12825.7–12831.7	E13	13018.5–13024.5	F13	13078.5–13084.5

(continued)

Table 3-3

CARS Band Frequencies (continued)

CATV Channel	VHF Boundaries (in MHz)	CARS Group C		CARS Group D		CARS Group E		CARS Group F	
		FCC Designation	Microwave Boundaries (in MHz)	FCC Designation	Microwave Boundaries (in MHz)	FCC Designation	Microwave Boundaries (in MHz)	FCC Designation	Microwave Boundaries (in MHz)
B	126–132	C14	12772.5–12778.5	D14	12831.7–12837.7	E14	13024.5–13030.5	F14	13084.5–13090.5
C	132–138	C15	12778.5–12784.5	D15	12837.7–12843.7	E15	13030.5–13036.5	F15	13090.5–13096.5
D	138–144	C16	12784.5–12790.5	D16	12843.7–12849.7	E16	13036.5–13042.5	F16	13096.5–13102.5
E	144–150	C17	12790.5–12796.5	D17	12849.7–12855.7	E17	13042.5–13048.5	F17	13102.5–13108.5
F	150–156	C18	12796.5–12802.5	D18	12855.7–12861.7	E18	13048.5–13054.5	F18	13108.5–13114.5
G	156–162	C19	12802.5–12808.5	D19	12861.7–12867.7	E19	13054.5–13060.5	F19	13114.5–13120.5
H	162–168	C20	12808.5–12814.5	D20	12867.7–12973.7	E20	13060.5–13066.5	F20	13120.5–13126.5
I	168–174	C21	12814.5–12820.5	D21	12873.7–12879.7	E21	13066.5–13072.5	F21	13126.5–13132.5
7	174–180	C22	12820.5–12826.5	D22	12879.7–12885.7	E22	13072.5–13078.5	F22	13132.5–13138.5
8	180–186	C23	12826.5–12832.5	D23	12885.7–12891.7	E23	13078.5–13084.5	F23	13138.5–13144.5
9	186–192	C24	12832.5–12838.5	D24	12891.7–12897.7	E24	13084.5–13090.5	F24	13144.5–13150.5
10	192–198	C25	12838.5–12844.5	D25	12897.7–12903.7	E25	13090.5–13096.5	F25	13150.5–13156.5
11	198–204	C26	12844.5–12850.5	D26	12903.7–12909.7	E26	13096.5–13102.5	F26	13156.5–13162.5
12	204–210	C27	12850.5–12856.5	D27	12909.7–12915.7	E27	13102.5–13108.5	F27	13162.5–13168.5

AML Frequencies (Bandwidth – 6 MHz)

	VHF	C		D		E		F	
13	210–216	C28	12856.5–12862.5	D28	12915.7–12921.7	E28	13108.5–13114.5	F28	13168.5–13174.5
J	216–222	C29	12862.5–12868.5	D29	12921.7–12927.7	E29	13114.5–13120.5	F29	13174.5–13180.5
K	222–228	C30	12868.5–12874.5	D30	12927.7–12933.7	E30	13120.5–13126.5	F30	13180.5–13186.5
L	228–234	C31	12874.5–12880.5	D31	12933.7–12939.7	E31	13126.5–13132.5	F31	13186.5–13192.5
M	234–240	C32	12880.5–12886.5	D32	12939.7–12945.7	E32	13132.5–13138.5	F32	13192.5–13196.5
N	240–246	C33	12886.5–12892.5	D33	12945.7–13951.7	E33	13138.5–13144.5		
O	246–252	C34	12892.5–12898.5	D34	12951.7–12957.7	E34	13144.5–13150.5		
P	252–258	C35	12898.5–12904.5	D35	12957.7–12963.7	E35	13150.1–53156.5		
Q	258–264	C36	12904.5–12910.5	D36	12963.7–12969.7	E36	13156.5–13162.5		
R	264–270	C37	12910.5–12916.5	D37	12969.7–12975.7	E37	13162.5–13168.5		
S	270–276	C38	12916.5–12922.5	D38	12975.7–12981.7	E38	13168.5–13174.5		
T	276–282	C39	12922.5–12928.5	D39	12981.7–12987.7	E39	13174.5–13180.5		
U	282–288	C40	12928.5–12934.5	D40	12987.7–12993.7	E40	13180.5–13186.5		
V	288–294	C41	12934.5–12940.5	D41	12993.7–12999.7	E41	13186.5–13192.5		
W	294–300	C42	12940.5–12946.5	D42	12999.7–13005.7	E42	13192.5–13198.5		

NOTE: Group C channels add 12,646.5 MHz to VHF Frequency; Group D channels add 12,705.7 MHz to VHF Frequency; Group E channels add 12,898.5 MHz to VHF Frequency; Group F channels add 12,958.5 MHz to VHF Frequency.

(continued)

Table 3-3

CARS Band
Frequencies
(continued)

FM Frequencies (Bandwidth = 25 MHz FM A, FM B; Bandwidth = 12.5 MHz FM K)

CARS group FM A		CARS group FM B		CARS group FM K			
Designation	Microwave Boundaries (in MHz)	Designation	Microwave Boundaries (in MHz)	Designation	Microwave Boundaries (in MHz)	Designation	Microwave Boundaries (in MHz)
A01	12700.0–12725.0	B01	12712.5–12737.5	K01	12700.0–12712.5	K21	12950.0–12962.5
A02	12725.0–12750.0	B02	12737.5–12762.5	K02	12712.5–12725.0	K22	12962.5–12975.0
A03	12750.0–12775.0	B03	12762.5–12787.5	K03	12725.0–12737.5	K23	12975.0–12987.5
A04	12775.0–12800.0	B04	12787.5–12812.5	K04	12737.5–12750.0	K24	12987.5–13000.0
A05	12800.0–12825.0	B05	12812.5–12837.5	K05	12750.0–12762.5	K25	13000.0–13012.5
A06	12825.0–12850.0	B06	12837.5–12862.5	K06	12762.5–12775.0	K26	13012.5–13025.0
A07	12850.0–12875.0	B07	12862.5–12887.5	K07	12775.0–12787.5	K27	13025.0–13037.5
A08	12875.0–12900.0	B08	12887.5–12912.5	K08	12787.5–12800.0	K28	13037.5–13050.0

A09	12900.0–12925.0	B09	12912.5–12937.5	K09	12800.0–12812.5	K29	13050.0–13062.5
A10	12925.0–12950.0	B10	12937.5–12962.5	K10	12812.5–12825.0	K30	13062.5–13075.0
A11	12950.0–12975.0	B11	12962.5–12987.5	K11	12825.0–12837.5	K31	13075.0–13087.5
A12	12975.0–13000.0	B12	12987.5–13012.5	K12	12837.5–12850.0	K32	13087.5–13100.0
A13	13000.0–13025.0	B13	13012.5–13037.5	K13	12850.0–12862.5	K33	13100.0–13112.5
A14	13025.0–13050.0	B14	13037.5–13062.5	K14	12862.5–12875.0	K34	13112.5–13125.0
A15	13050.0–13075.0	B15	13062.5–13087.5	K15	12875.0–12887.5	K35	13125.0–13137.5
A16	13075.0–13100.0	B16	13087.5–13112.5	K16	12887.5–12900.0	K36	13137.5–13150.0
A17	13100.0–13125.0	B17	13112.5–13137.5	K17	12900.0–12912.5	K37	13150.0–13162.5
A18	13125.0–13150.0	B18	13137.5–13162.5	K18	12912.5–12925.0	K38	13162.5–13175.0
A19	13150.0–13175.0	B19	13162.5–13187.5	K19	12925.0–12937.5	K39	13175.0–13187.5
A20	13175.0–13200.0			K20	12937.5–12950.0	K40	13187.5–13200.0

Figure 3-24
Microwave path
profile

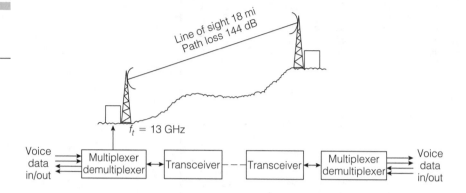

groups of frequencies: C, D, E, and F. Note that group F can handle only 32 VHF carriers (2-M). This AML system used by cable operators is called community-antenna radio service (CARS) band and is in the frequency band of 12.7 to 12.95 GHz.

Of the groups in Table 3-3 from D33 to F32 use is specified for studio transmitter link (STL) service. This AML service is used for connecting the studio facilities, usually in a city, to the transmitter or uplink site outside of the city. The antenna systems used at these frequencies are usually dish type (parabolic) from 4 to 10 ft in diameter. The transmitting antenna may be 10 ft, and the receiving antenna size may be 6 ft in diameter. As in satellite antennas, the larger the diameter the greater the antenna gain. A diagram of the basics is shown in Figure 3-24. Now, several other factors that add to this loss have to be considered.

3.52 Telephone–Microwave Radio Link Applications

The microwave-radio link has been used by the telephone industry to transmit and receive telephone traffic for many years. Early systems used the carrier method of single-sideband suppressed-carrier modulation for transmission of a one-way (half-duplex) telephone call. A carrier operating on another carrier going the other direction carried the other half of the call. This technique is known as frequency division multiplexing (FDM). Later, telephone voice signals were converted to nonreturn to zero (NRZ) digital pulse streams, referred to as time-division multiplexing (TDM). Both methods were used to stack voice channels on one RF microwave carrier. This carrier is a major concern and should be received at the receiving site in the best possible condition. The basic microwave link for one path is shown in Figure 3-24.

Path length L = 18 miles path loss

$$= (36.6 + 20 \log L + 20 \log f_t) \text{ dB}$$

$$= 36.6 + 20 \log 18 + 20 \log 13{,}000$$

$$= 36.6 + 25.1 + 82.3$$

$$= 143.98$$

$$= 144 \text{ dB}$$

This loss assumes nothing is in the pathway and there is no signal refraction.

3.521 The wave front from the transmitting antenna expands as it travels through space, which results in reflections and phase changes as the wave passes over obstacles and obstructions. Extra height of the transmitting and receiving sites should be allowed. The beam cross section shows first, second, and third Fresnel zones as concentric bands around the beam center axis, as shown in Figure 3-25.

These formulas are used to adjust tower heights and site elevations so objects or obstructions do not penetrate the first Fresnel zone.

3.522 The procedure of path profiling is to make sure that the transmitting and receiving site is clear of obstructions and that the path or beam, taking into account the Fresnel zones, is far enough above the Earth's

Figure 3-25
Fresnel zones

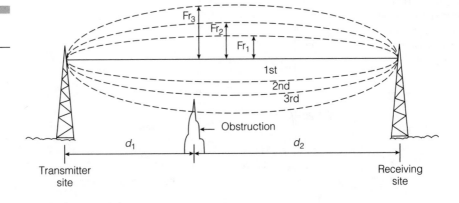

surface to be unaffected. This procedure might require some on-site work with surveying instruments. However, reasonable estimates can be made using 7.5-minute topographical maps and appropriate graph paper. Also, sea-level refractivity contour charts as well as K factor versus refractivity might be helpful in estimating the K factor for the area. A short example illustrates this procedure as shown in Figure 3-26.

Example 3-4

To correct this figure for points *x* and *y*, we can use the formula or the graph in Figure 3-27.

At point *y*, Fresnel radius is 0.90 of mid-span value of 42.4 ft. So, 0.90 × 42.4 ft = 38 ft. At this point, the bulge of the Earth penetrates

Figure 3-26
Complete path profile

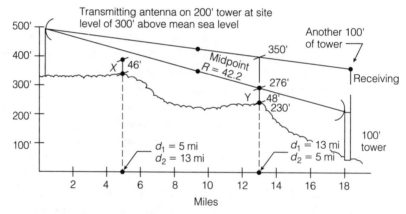

Figure 3-27
Conversion from mid-path Fresnel zone clearance for other than mid-band obstacles

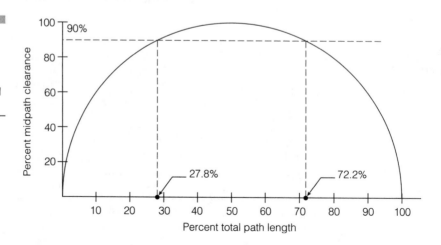

the Fresnel zone, indicating no clearance. Another 100 feet of receiving tower height should do the job. At y, h = 46 ft + 230 ft = 276 ft. The corrected Fresnel radius = 276 ft + 38 ft = 314 ft + path line of site elevation at point y = 350 ft, so there is proper clearance allowances for vegetation. Trees might make an additional 50 ft of receiving tower height more prudent.

Once the microwave path is cleared, recall that the 18-mile path loss was 144 dB = PL. Suppose this was a high-power transmitter with an output power of 1 W converted to dBm. Recall that 1 mW corresponds to 0 dBm.

$$P_T = 10 \log \frac{1 \text{ W}}{1 \text{ mW}} = 30 \text{ dBm}$$

If the gain of the transmitting antenna is 45 dB, then the radiated power P_R is calculated.

$$P_R = 30 \text{ dBm} + 45 \text{ dB} = 75 \text{ dBm}$$

At the receiving antenna we may calculate,

$$P_{\text{REC}} = P_R - P_L = 75 - 144 = -69 \text{ dBm}$$

If the gain of the receiving antenna is 40 dB, then the receiving power P_{REC} is calculated as

$$P_{\text{REC}} = -69 + 40 = -29 \text{ dBm}$$

If the receiver manufacturer specifies the noise power level as -90 dBm, then the receiver input carrier-to-noise ratio is calculated as

$$C/N = C - N = -29 - (-90) = 59 \text{ dB} = 60 \text{ dB}$$

where C and N are measured in dB. The carrier level is 60 dB greater than the noise level.

A good receiver should be able to produce a base-band S/N or about 52 dB with a 60 dB carrier-to-noise ratio specification.

The fade probability of an outage occurring based on Rayleigh fading the worst month of the year can be calculated by

$$P_F = T_F \times T_{TF} \times 2.5 \times 10^{-6} \times f \times L^3 \times 10^{-F_M/10}$$

P_F = fade probability of an outage occurring

T_F = terrain factor
4 for smooth terrain and water
1 for average terrain
0.25 for rocky, mountainous terrain

T_{TF} = temperature/humidity factor
0.5 for hot or humid areas
0.25 for normal climate
0.125 for high-mountain and dry areas

f = operating frequency in GHz
L = length of path in miles
F_M = system fade margin 35 dB AML, 30 dB others

Example: Normal terrain and normal climate for the 18-mile hop at 13 GHz and 30-dB fade margin.

$$P_F = 1 \times 0.25 \times 2.5 \times 10^{-6} \times 13 \times 18^3 \times 10^{-30/10}$$

$$= 8.1 \times 10^{-6} \times 18^3 \times 10^{-3}$$

$$= 8.1 \times 10^{-6} \times 5832 \times 0.001$$

$$= 47{,}239 \times 10^{-6} \times 0.001$$

$$= 0.047239 \times 10^{-3}$$

For a month consisting of 720 hours, it amounts to

$$0.0472 \times 10^{-3} \times 720 \text{ hr} = 30.5 \times 10^{-3} \text{ hr}$$

$$= 0.031 \text{ hr}$$

$$= 0.031 \text{ hr} \times 60 \text{ min/hr}$$

$$= 1.86 \text{ minutes in a month}$$

Heavy rain can affect the line-of-sight microwave link. At frequencies below 10 GHz, rain does not have much effect. Studies of rainfall over a five-year period can indicate seasonal rains that could cause problems.

Summary

At the headend point in a cable system, the signals that make up the system's product are assembled and grouped according to the channel lineup. This channel lineup or grouping is usually the result of an agreement between the cable operator and the licensing or franchising municipal authority. The signal quality at this point in the system is as good as it's ever going to be. It's the job of the distribution system to deliver this product as perfectly as possible to the system subscriber. At the headend the electronic equipment cleans up and adjusts the signal level on a channel-by-channel basis. Off-air signal processors remove signal level variations and filter unwanted noise and interference products. An Earth receiving antenna and receiver set picks up satellite programming and reduces the resulting signal to its base band. This base-band signal is the usual NTSC video and sound signal or a digital bit stream. In most cases this digital signal is decoded further into video and audio signals, which then are converted to standard NTSC signal for subscribers' television sets.

More people in the television viewing public are purchasing large-screen HDTV sets with a 16:9 aspect ratio. Such sets usually offer a standard NTSC tuner as well as a digital HDTV tuner; these tuners allow the subscriber to view both types of signals. The satellite signals that appear as standard NTSC are modulated onto a standard television channel and connected to the cable system. Digital signals are usually in the standard MPEG-2 or MPEG-4 format and are carried by the cable systems using digital modulators of the QAM 64 or 256 format. Most subscribers with digital TV sets have the capability to receive in the digital mode the broadcast station format of eight-level vestigial sideband (8-VSB) or the QAM 64 or 256 format.

At the headend all the channels are combined down to a single point needed to drive the cable trunk system. All signals from the broadcast networks, satellite systems, microwave systems, as well as any locally generated programming enter the cable trunk system from the headend.

Many cable operators have some form of bidirectional operation signals that arrive from many points in the system back to the headend. Cable operation in the reverse direction is usually in the 5 to 30 MHz reverse band, which is below the lowest television carrier in the forward direction, which is television channel 2. At the headend the signals are received, processed, and delivered to the appropriate destination.

Here the control of the cable system exists, hence the building housing the headend equipment should be closely environmentally controlled and monitored. In most instances the headend building is surrounded by a security fence and has an intrusion alarm as well as temperature and signal monitors. Usually, headends employ stand-by emergency power systems. Many cable operators have telemetering links connecting the company offices with monitored information.

Questions

1. Explain the reason that antenna spacing on a television-receiving tower is critical.
2. Explain why a preamplifier may be needed for a tower-mounted antenna.
3. For an antenna tower system, what must be done to protect the headend from the effects of lightning?
4. Describe the emergency alert system (EAS) for a cable operator.
5. Describe the basic purpose of a video/audio television modulator.
6. For a satellite receive-only station, explain why it is advisable to have frequency down-conversion at the antenna feed point.
7. What is the main reason to harmonically relate the cable carrier frequencies at the headend?
8. List two possible locations for an off-air preamplifier at the receiving-antenna tower site.
9. Explain the purpose of the headend combining network.
10. Describe the difference between a signal processor and what is called a demod/remod scheme.

Problems

1. Calculate the minimum vertical separation between two antennas mounted one below the other on the same tower leg, if the lowest frequency is television channel 7.
2. Calculate the minimum spacing in height between a television channel 7 antenna if the additional antenna is being placed at right angles to the television channel 7 antenna.

3. An eight-input combining network provides a constant +22 dBmV output to the trunk feed point. If this combining network is constructed of 8 dB directional couplers with a 1 dB through loss, calculate the eight input levels.

4. If two-way splitters are going to be used as an eight-input combiner, how many of these devices will be needed?

5. For the combining network in problem 4, calculate the input levels for a constant output level of +22 dBmV. Assume each splitter is perfect with a loss of 3 dB.

6. If a microwave path is to be operated at 13 GHz and the hop distance is 30 mi, calculate the path loss.

Fiber-Optic Technology in Cable Systems

Objectives

After learning the material in this chapter the student will be able to

- Explain the development of optical fiber as a communication method.

- Describe optical light sources and detectors as the light transmitters and receivers.

- Describe the various methods of splicing and connectorizing optical fibers.

- Identify the areas of fiber-optical applications in a cable television system.

- Explain the construction practices required to properly install fiber-optical cables.

- Describe proper handling of fiber-optical cable and recommended preliminary admittance testing procedure.

- Explain the final test and acceptance procedure following construction.

4.1 Introduction

Fiber-optic technology as applied to cable television systems was just what the industry needed. Cable systems were expanding their bandwidth with the upper limit approaching 1 GHz. The number of amplifiers per cable-mile of plant was increasing at an alarming rate. High power consumption required more power supplies, which in turn caused system costs to increase. More cable plant devices and connectors caused increased signal leakage problems, all contributing to increased plant maintenance and escalating costs. It was indeed a difficult and expensive project to expand plant bandwidth and channel capacity. Fiber-optic technology started out as simply plastic light pipes acting as monitor indicators for a variety of applications including automobiles. The continued research into the field of optics and optical fibers finally produced fiber-optic communications.

4.11 Fiber-Optic Development

The development of optical fiber technology began in the 1950s and 1960s with the invention of light amplification by a stimulated emission radiography device, the laser. Early lasers were essentially research tools used to study transmission of light through various optical devices, glass fiber included. The first lasers were gas-type lasers with large amounts of light energy transmitted. The manufacturing methods for forming glass (silica) fibers were developed by companies in the United States and abroad. As light sources were developed, glass fibers contributed to the development of fiber-optic communications.

4.111   The significance of the laser was due to its capability to supply a high-intensity light output at essentially one wavelength and be coherent (in phase) as well. Thus, it was described by its action as a monochromatic coherent light source. The laser of choice for communication applications became the solid-state laser diode. This device was small enough to be mated with the end of an optical fiber. The development of the photodiode was also important to the communication application because this was the photo detector or receiver. This photodiode was the transducer that converted the received light energy to electrical energy. Now the ingredients—the transmitter, the fiber-optic cable, and the receiver—were available for a fiber-optic communication system.

4.112   A review of applicable optical physics will help give more of an insight into how an optical fiber communications system operates. For a fiber-optic communications link, the carrier is optical energy that propagates through the fiber to the receive point. The optical transmitter at the sending end has to be modulated with the electrical signal. At the receiving end, the receiver converts the modulated optic wave to electrical energy that is nearly an exact replica of the sending end electrical signal. The transmission medium from the sending end to the receiving end is a glass fiber.

Two types of silica glass fibers exist; one is called multimode and the other single mode. The modes are essentially the same as waveguide modes in the RF domain. We should recall that radio frequency (electromagnetic) radiation travels at the speed of light (in a vacuum and nearly so for air). This is because light energy is electromagnetic and propagates through space as well as glass fibers. It should come as no shock that glass fibers operate as waveguides for light energy. Often when light energy is mentioned, we think of visual light. If a person looks through a

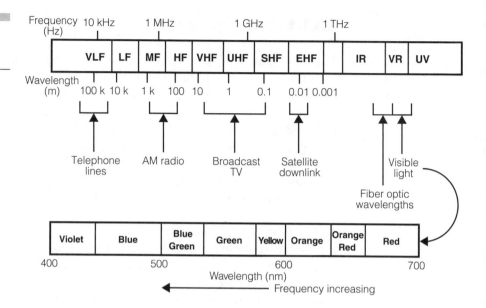

Figure 4-1
Electromagnetic spectrum

thick sheet of window glass at the edge, the light generally looks greenish and is reduced in brightness. Therefore, this type of glass would not work very well if used to form a fiber. In brief, the glass has to be extremely clear and transparent to light energy. Glass manufacturers have discovered that glass transparency is a function of color or wavelength. Because optical energy is electromagnetic, a spectrum of electromagnetic energy is shown in Figure 4-1.

The visual light spectrum, with its associated colors, occupies an extremely small portion of the frequency spectrum. The plot of an optical silica glass fiber is illustrated in Figure 4-2. Notice that the visible spectrum is to the left of the graph (less than 1000 nm). Also notice for a wavelength of 1550 nm and 1310 nm, the loss is 0.2 dB/km and 0.3 dB/km. This simply means that energy at 1550 nm of wavelength will be decreased only 0.2 dB for 1 km. A coaxial cable with one of the lowest losses at 1000 MHz is about 1.3 dB/ 100 ft at 68° F. This is approximately 43 dB/km. It is this extremely low loss that makes the use of optical fiber attractive in the communications industry.

4.113 The first generation of optical fibers were a little more than light pipes used as indicating devices, such as automotive lamp monitors and inspection monitoring equipment. As research and development continued, it was found that if the glass fiber was coated with glass of a slightly lower refractive index, total internal reflection would take place. Thus,

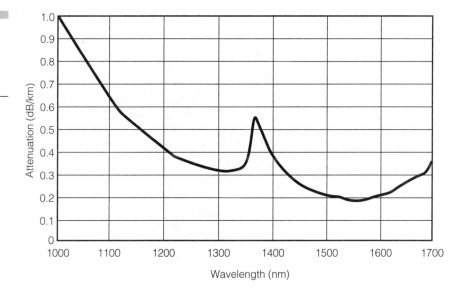

Figure 4-2

Attenuation versus wavelength for a glass single-mode fiber

light energy entering the fiber would stay in the core and travel to the end of the fiber. This outer covering of glass is known as the cladding. Light rays essentially propagated through the fiber core in waveguide fashion, following a multitude of mode paths, and multimode optical fiber resulted.

Variations in the geometry of the core-cladding cross section at the end of a fiber resulted in several sizes. When the refractive index of the cladding is slightly less than that of the core and the change is abrupt, the resulting optical fiber is known as the step index. This means that there is a step change in refractive index at the core cladding boundary; such optical fibers are known as step index multimode fibers. This type of optical fiber is shown in Figure 4-3. Rays of light energy propagate through the fiber following different paths due to the different modes of operation, as shown in Figure 4-4.

Due to the different path lengths, light rays at the end of a fiber arrive at different times. This causes a sharp input light pulse to become stretched in time at the output. This stretching in time is known as modal dispersion. Modal dispersion can become a limiting factor for a high pulse rate found in digital signal transmission.

Early fiber-optic systems were used in the local area networks (LAN) with data (digital) communications connecting computer workstations to printers and file servers. Since many of these systems did not consist of long fiber runs or excessively high pulse (data) rates, system performance

Figure 4-3

Cross section of step-index multimode fiber

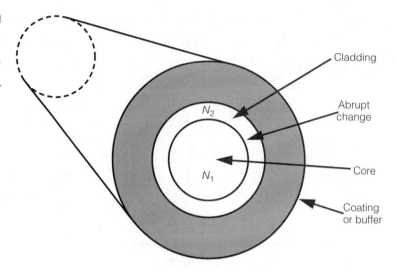

N_1 = Refractive index of core glass
N_2 = Refractive index of cladding glass
$N_1 > N_2$ by approximately 1%

Figure 4-4

Multimode light-wave propagation through fiber

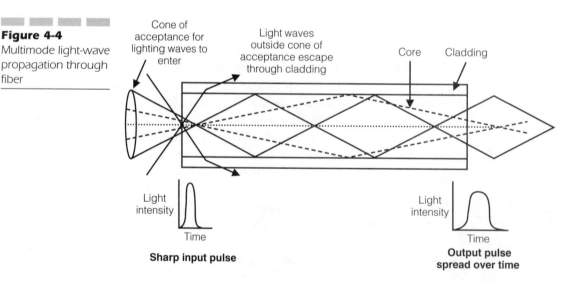

Sharp input pulse

Output pulse spread over time

Note: Multimode propagation causes output-pulse dispersion.

was adequate. As system requirements called for greater distances and higher data transfer rates, modal dispersion became a limiting factor. The industry responded with an improved multimode optical fiber known as a graded-index fiber. A plot of the magnitude of refractive index of the glass versus the fiber diameter is shown in Figure 4-5 for both step-index and graded-index multimode fibers.

Notice in the figure that the refractive index of the glass for the graded-index optical fiber decreases gradually from the core center to the outside cladding. Fiber manufactured in this manner allows the rays of light entering the fiber to propagate through the fiber, as shown in Figure 4-6. Light rays travel faster in the cladding area where the refractive index

Figure 4-5

Refractive index of types of optical fibers

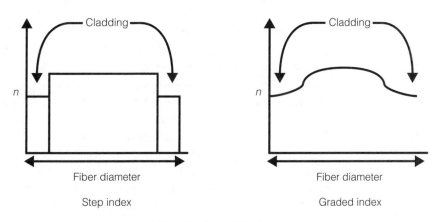

n is the refractive index of the glass

Figure 4-6

Graded-index multimode fiber

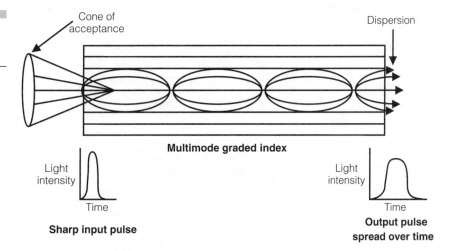

is less than the core center. Therefore, they may travel a longer path yet travel faster to arrive at the end along with rays traveling different paths. The effects on pulsed signals are not as severe as with the step-index multimode fiber.

In general, multimode fibers have a relationship between the refractive index of the core and cladding glass and the numerical aperture, as well as the acceptance angle. This relationship is illustrated in Figure 4-7 for a step-index multimode fiber.

For cable television purposes, long distances of a fiber operating with wide bandwidth signals are required, and neither of the multimode fibers is applicable. Most multimode systems can operate quite well at shorter distances using light-emitting diode (LED) technology as the transmitter and a photodiode as the receiver. The larger core diameter with the larger numerical apertures allows the use of these diodes. The fiber of choice for high-speed wideband fiber systems required the development of a single-mode optical fiber. This fiber has a very small core diameter and a thick cladding surrounding it. The small core diameter makes it more difficult to get much optical energy into the fiber.

When the solid-state laser chip was developed, it made it possible to use single-mode fiber systems. The laser chips were made small, and the light-emitting area was matched closely with the end of the fiber. The high-intensity narrowband coherent laser light source was just the component needed for single-mode fiber operation. The most common single-mode fiber used today is step index and has a very small core cross section. Figure 4-8 compares the cross-sectional areas for multimode and single-mode fibers.

Figure 4-7

Fiber-optic geometry

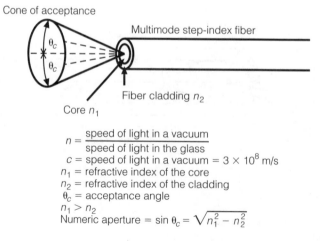

Cone of acceptance

Multimode step-index fiber

θ_c

θ_c

Fiber cladding n_2

Core n_1

$n = \dfrac{\text{speed of light in a vacuum}}{\text{speed of light in the glass}}$

c = speed of light in a vacuum = 3×10^8 m/s

n_1 = refractive index of the core

n_2 = refractive index of the cladding

θ_c = acceptance angle

$n_1 > n_2$

Numeric aperture = $\sin \theta_c = \sqrt{n_1^2 - n_2^2}$

Figure 4-8
Fiber cross sections

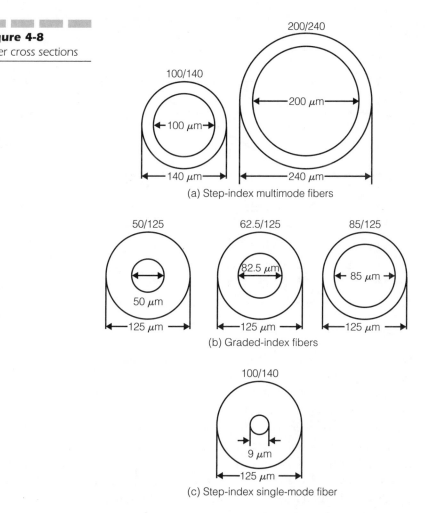

(a) Step-index multimode fibers

(b) Graded-index fibers

(c) Step-index single-mode fiber

For the single-mode fiber shown in the figure, it should be noted that the outside diameter is the same as for the graded-index multimode fiber. This illustrates that the outside diameter says nothing about the fibers' characteristics. The actual fibers making up commercially produced fiber-optic cables are coated with a colored plastic covering. A color is used to identify each fiber placed in a protective buffer tube.

A single-mode optical fiber provides low attenuation of light as it propagates through the fiber, and because modal dispersion is not a factor, very high pulse transmission rates can be used. As the upper frequency and pulse transmission rates are increased, some dispersion does result. The remaining dispersion in a single-mode fiber is waveguide dispersion and

Figure 4-9
Chromatic dispersion
versus wavelength

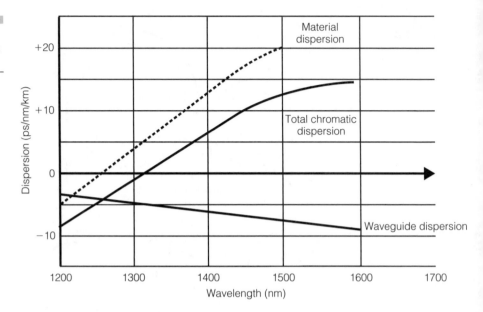

material dispersion. Both types vary as a function of wavelength and are collectively called chromatic (color) dispersion. Waveguide dispersion is negative and material dispersion is both negative and positive. A graph of these two dispersions and the combination are shown in Figure 4-9.

Examining this figure shows that the effects of waveguide dispersion and material dispersion cancel at approximately 1310 nm of optical wavelength. This results in zero dispersion, so little or no pulse stretching (distortion), and high optical pulse rates can be used. After examining Figure 4-2, it should be noted that the fiber loss at 1310 nm is not at the lowest point. The loss is at its lowest at 1550-nm. In Figure 4-9, we can see the dispersion is not zero at 1550, which means high data rate pulsed signals will be affected. The development of doped fiber results in shifting the zero dispersion point into the 1550 nm region. Now high data rates at the lowest loss area for this doped fiber become possible. This is the fiber of choice for long-haul optical cable systems such as submarine cables.

4.12 Light Sources and Their Development

Operation of an optical fiber communication system starts at the light transmitter, which injects the optical signal into the fiber. This light source has to be modulated with the message signal. The stronger or more intense the light source becomes, the stronger the received signal. Light energy traveling through the optical waveguide carries the message traffic to the receiving end. The light source at the transmitting end has to be modulated with the message-carrying signal.

The light sources used in current systems are either a solid-state LED or a solid-state laser diode. LEDs are often used for LANs carrying multimode data and provide satisfactory operation for a variety of applications. The solid-state laser diode is used mainly in single-mode applications, requiring communication distances of 10 to 50 km. This is the type used in cable television systems. There are two basic types of laser diodes: the Fabry-Perot and the distributed feed back (DFB). The characteristics of these diodes differ, as shown in their characteristic curves in Figure 4-10.

Since the laser bandwidth is important in the calculation of the effects of dispersion in fiber-optic systems, it may not seem important to discuss LED light sources given that the solid-state laser diode is the optical transmitter of choice for cable television systems. Many buildings, however, are already wired using multimode fiber. In many instances, apartment buildings as well as commercial office buildings contain unused optical fiber. Such fiber could be used to connect services to subscribers because many of the distances are short compared to ordinary cable systems' fiber-optic plant. Small multimode optical detectors could feed the RF signal spectrum to a mini-distribution coax system or directly into a converter.

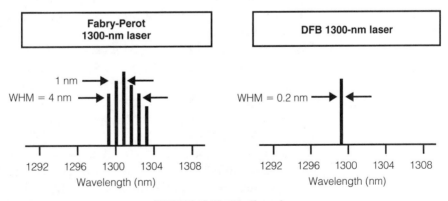

Figure 4-10
Laser spectrums

4.121 Solid-state LEDs were used in many early type data communication optical-fiber systems. Much was learned about these light sources and many improvements were made. LEDs had several problems associated with their use as optical transmitters. The first was speed. An LED cannot be modulated at rates (speed) as high as a laser. Also, LEDs do not have as high an optical output as lasers do.

Because speed and intensity are important factors for light sources acting as transmitters, lasers are the best choice. Generally, LEDs have a much wider spectral width than lasers, and this increases the effects of modal dispersion in multimode systems. Dispersion causes the stretching of short pulse signals in time, which severely limits the pulse or data rate if it becomes excessive. LEDs are made to operate in the 850-nm and 1300-nm regions, which are in the lower operating loss regions for glass optical fiber.

4.122 The solid-state laser diodes used in most single-mode, long-distance fiber-optic systems can be made to operate in all the low loss optical windows of most fibers. The 1300-nm and 1500-nm regions are used in cable television single-mode fiber-optical systems. Both light-emitting diodes and laser diodes are directly modulated by control of the drive current. Both have operating characteristics that vary with temperature. LEDs in general can operate within the range of $-25°$ C to $+125°$ C. Lasers are somewhat more temperature sensitive. Therefore, cooling using a Peltier cooler may be required in some applications. When dealing with laser diodes in cable television systems operating in the usual headend air-conditioned environment, proper heat sinking is adequate. A typical operating curve of optical output versus drive current is shown in Figure 4-11.

An examination of this figure indicates that the operating region is the linear portion of the curves for both the LED and the laser. Notice that the temperature characteristics are different and the LED is operational over a wide temperature range. The extreme linearity of the laser diode and the high optical power output are important factors that make the laser a superior transmitting device.

A laser diode light source is available as a single device or as a packaged unit mounted in a heat-sink enclosure. Most laser diodes in cable television applications appear in system optical transmitters that include the associated stabilization and drive circuits. The RF signal, consisting of the band or bands of cable television channels, is connected to the input. The optical transmitter converts this signal to a modulated light signal that is coupled to the fiber-optic transmission cable. The transmitter out-

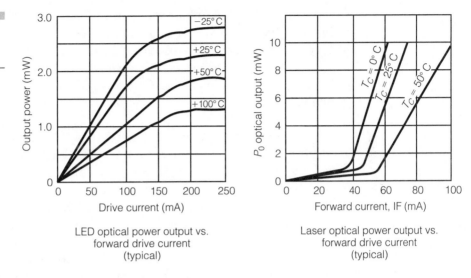

Figure 4-11
LED and laser
characteristic curves

LED optical power output vs.
forward drive current
(typical)

Laser optical power output vs.
forward drive current
(typical)

put can be divided by a fiber coupler, which is connected to several fibers feeding different areas of the system.

4.13 Optical Detectors

It is equally important that the received optical signal at the receive point be accurately detected and connected to the cable distribution system. A variety of photodiodes with a multitude of characteristics have to be considered in making a choice. One of the more important characteristics is responsiveness. In effect, this is the sensitivity for the photodiode. Responsiveness is given in amperes per watt. The watt figure is the optical power falling on the photodiode's junction, which produces a current through the photodiode.

Rise time is another important parameter. The rise time is the time it takes for the current to reach its stable value as a sudden increase in light falls on the diode junction. Photo diodes are reverse-biased so the electrical signal is essentially the current through the junction, causing the diode's resistance to change. When no light falls on the photosensitive diode junction, the resulting current is called the dark current.

4.131 Optical detectors have gone through many improvements. Early detectors were essentially photocells that would produce a small voltage that would vary with changes in light levels. Some people might recall the cadmium sulfide photocell that varied its resistance with changes in light levels. The silicon solar cell produces an electrical voltage when light rays fall on the cell's photosensitive area. Most of the photo or light-sensitive devices previously discussed either require significant amounts of light or are too slow to be of much value in fiber-optic communication applications.

4.132   Solid-state photodiodes are used as the light detector or receiver in fiber-optic communication systems. The photodiode's p-n junction is reverse-biased, which removes current carriers out of the depletion region that blocks current until light enters that frees electron's hole pairs, causing the current to flow through the junction. Doping can change the dark current of the photodiode and thus low light level operation can be improved. Since the transistor consists of either an n-p-n or p-n-p junction, if one of the junctions, usually the base emitter junction, becomes the photodiode, the photo transistor results. Furthermore, if the photo-transistor is placed with a second transistor into a Darlington circuit, the Darlington photo detector results.

As might be suspected, all of the gain results in an increase in responsiveness. Again, it might seem that this type of light detector would indeed be superior, but the junction capacitance causes the rise time to increase. This, of course, affects the high-speed pulse (data) response and high-frequency response. Phototransistor detectors are ideal for high sensitivity and medium-to-low speed applications.

4.133   The pin-photodiode does not have the responsiveness of amplified-photo detectors, but it has a very small rise time and can be used for high-speed, high-frequency applications. The pin-photodiode is usually used for single-mode operations for cable television optical receivers. Usually, an optical receiver circuit uses an operational amplifier following the pin-photodiode, which does not load the photodiode output and provides some power gain as well as a low impedance output to the following circuitry. Companies that manufacture optical transmitters and receivers for cable television applications usually offer optical transmitters with input levels and a number of cable channels specified. Optical receivers are usually paired with the optical transmitter.

The cable operator has to design for a given power budget calculated by the system losses, the transmitter output, and the receiver threshold sensitivity. An example of an optical transmitter and receiver fundamental circuit is shown in Figure 4-12. A listing of various photodiode detector characteristics is shown in Table 4-1.

Figure 4-12a
Optical transmitter

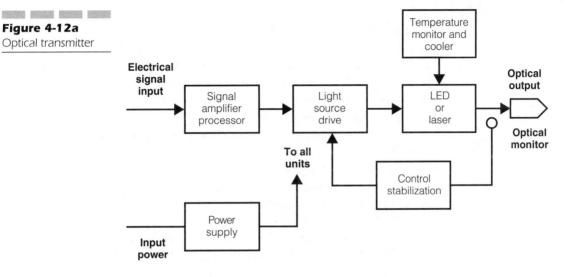

Figure 4-12b
Optical receiver

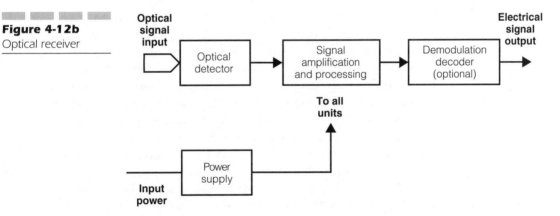

Table 4-1

Photodiode
Detector
Characteristics

Device Name	Dark Current (in nA)	Rise Time (in ns)	Responsivity (in A/W)
(Si) pin-photodiode	10.0	0.1–5	0.5
(InGaAs) pin-photodiode	0.1–3	5 ps–5	0.75
(Ge) avalanche photodiode	400	0.25–1	0.6
(InGaAs) avalanche photodiode	30	0.3	0.75

4.14 Splicing and Connectorizing

Splicing and connectorizing optical fiber into a network is crucial to the operation of a fiber-optic communication system. Since the optical fiber exhibits such low loss over a distance, a system should not allow losses to accumulate from the necessary splices and connectors. Prudent designers should order sufficient optical cable for segments that require a branching or termination point. Since connectors are required at stations in the system, extra splices along the way should be avoided. The quality of splicing and of connectors should be good so that the accompanying loss is held to a minimum. A variety of connectors are available for connecting a system. But a good design dictates that the variety of connectors should be minimized. This means that personnel will not have to be trained for different connectors. Also, stocking a variety of connectors and the necessary tooling can be avoided.

4.141 One of the first splices to be used was a mechanical splice. Essentially, the optical fiber was cut, cleaned, and cleaved so the end was cut flush (perpendicular to the axis). When both ends were thusly prepared, the ends were placed in a ferrule (tubing) made of either metal alloy or plastic and butted together end to end. Improvements in such types of splices included the addition of an index-matching gel to help mate the cleaned ends and a cement to prevent the fiber end from pulling.

Another big improvement in mechanical splices was what became known as the elastomeric splice. This type of splice used a plastic holder that held the splice open with a tool. The cleaved and cleaned ends of the fibers entered from opposite ends of the plastic splice piece, where they met at a junction point containing an index-matching gel. When the tool was withdrawn, the fibers in the grooved track clamped closed, holding the fibers at the junction. Properly made elastomeric splices often yield splice loss less than 1 dB. The splicing of optical fiber was always considered to be permanent even if such splices of the elastomeric type could be opened and the fibers disconnected.

4.142 The method of choice for splicing optical fibers for both multimode and single mode is called fusion splicing. This method essentially melts and fuses the glass fibers together. The equipment needed to fusion splice optical fibers is a sophisticated and relatively expensive device requiring proper operating expertise. Fusion splicers, as they are known, are made to splice a single fiber at a time. The new flat or ribbon fiber cable can be spliced several at a time.

Again, the procedure requires the fibers to be cut the prescribed length, have any plastic covering removed right down to the glass, and be cleaved using a diamond-wheeled cleaver. The ends are placed on the work stage of the splicer and are clamped in place by spring-loaded clamps. The fibers are essentially laid in "V"-type grooves on the work stage. An optical microscope allows the operator to move the ends to an end-to-end connection. An electric arc is turned on to raise the junction temperature to the melting point and then the arc is turned off. The junction of the two fibers becomes fused.

Present-day splicers place a sharp beam of light into one end of the splice joint while a photodetector at the other side detects light passing through the splice. This test allows the operator to measure the light loss through the splice. If the splice loss measured is too much, the operator can then redo the splice.

Fiber-optic splices for single-mode fibers are placed in protective tubing that contains a metal strength member. This protective tubing is the heat-shrink variety, and many fusion-splices have a heated drawer so this tubing can be shrunk down immediately after being spliced and tested. Spliced fibers in their respective tubing are clamped in splice trays containing extra, loose fiber placed in a figure-eight channel.

Testing of the splice loss is completed right after the fusion has cooled. A stress test—during which a slight tension is placed on the splice— is performed so the operator knows that fusion actually took place and the fibers weren't just placed together. Following this test, the protective sleeve is slid over the fused junction and is heat-shrunk down.

The optical viewing method has been improved, so most splicing equipment has a miniature television camera attached to the microscope and the image projected on an LCD screen. Some splicing gear even has a video output that allows the operator the capability to connect a large-screen monitor to the equipment. The monitor can be used in training other prospective splicers in the proper techniques. A rudimentary sketch of a fusion splicer is shown in Figure 4-13. Also, Figure 4-14 shows some examples of typical splicing termination equipment. The trays containing the splices have input and output ports for the fiber-optic cable to enter and exit the splice enclosure. The enclosures are usually the PVC plastic-sealed aerial kind or the plastic-sealed cabinet type for surface-mounted pedestals.

Figure 4-13
Optical fiber fusion
splicer

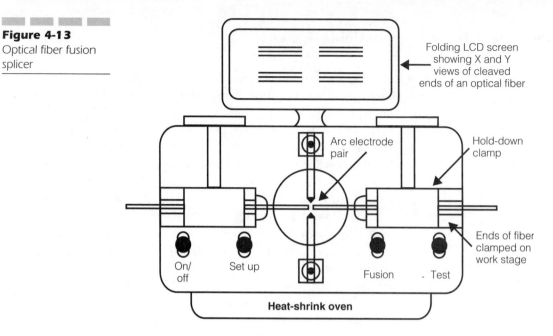

Folding LCD screen
showing X and Y
views of cleaved
ends of an optical fiber

Arc electrode
pair

Hold-down
clamp

Ends of fiber
clamped on
work stage

On/
off

Set up

Fusion

. Test

Heat-shrink oven

Figure 4-14
Splice termination

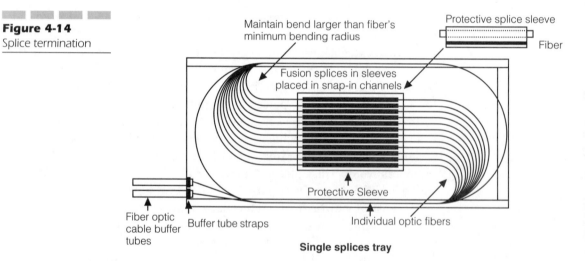

Maintain bend larger than fiber's
minimum bending radius

Protective splice sleeve

Fiber

Fusion splices in sleeves
placed in snap-in channels

Protective Sleeve

Fiber optic
cable buffer
tubes

Buffer tube straps

Individual optic fibers

Single splices tray

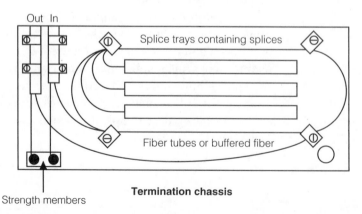

Out In

Splice trays containing splices

Fiber tubes or buffered fiber

Strength members

Termination chassis

4.143 As discussed earlier, several types of connectors are used in optical-fiber networks. For cable television systems using mostly single-mode fiber, three main types of connectors are used (see Figure 4-15). Optical connectors essentially allow the bare end of the fiber to terminate at the end of the connector. This end is polished after installation, which can be done manually with a polishing holder or mechanically using a power polisher.

The installation of connectors, like fusion splicing, requires sufficient tools and expertise. In many instances, short pieces of optical fiber with connectors installed on each end will be purchased. Such fiber pieces are known as jumper cables or pigtails. These jumpers are usually cut in the center and are fusion-spliced to the optical fiber. The optical fiber is then connectorized and is simply connected to the optical transmitter or receiver as required. Such a procedure avoids having a connector crew to connectorize a system, especially if not many connectors are needed. For systems requiring an optical patch panel, it is best to have it professionally connectorized.

In a typical case, fiber cables enter a termination station where the individual fibers in their color-coded buffer tubes are spliced to a single-fiber cable with a connector installed. These single connectors usually run through tracks or channels to the electronic equipment racks containing

Figure 4-15

Types of connectors
commonly used on
cable television fiber-
optic equipment

FC

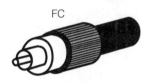

Typical insertion loss 0.5–1.0 dB
Push in and turn to lock connector

D4

Typical insertion loss 0.2–0.5 dB
Screw-type connector

SC

Typical insertion loss 0.2–0.5 dB
Push-pull-type connector

the optical receivers or transmitters. If this station is simply a receiving station, the optical signals are connected to the receivers, where optical to electrical (electronic) conversion takes place. If the station is a receive-and-transmit station, then communication services can essentially undergo drop or add and be transmitted to other stations.

4.144 Quite recently, the fiber-optic coupler or splitter has been perfected. This device is actually an optical fusion of more than just two optical fibers. Figure 4-16 illustrates the concept of splitting and coupling the optical signal. Because this device is made from optical fiber itself, it is quite small and physically compact. Such optical signal splitters can be placed in either aerial splice closures or aerial optical receivers. They can also be found in terminating stations where an optical signal enters at a high level, drops off a low level signal, and exits on another optical cable to another receiver station.

A device that is often needed is an optical attenuator used to reduce a signal into a receiver so as not to overdrive it. Optical attenuators are made by a calibrated bend in the optical fiber. Here, light escapes from the fiber right through the cladding and plastic coating. Lost light results in an attenuated signal into the device connector. The devices are constructed with a plastic form with a bending channel radius built in with the calibrated bend. The fiber is placed in the groove and the cover is closed. Another method of making an attenuator is with some fusion splicers. Here, the core of the joined fibers is offset an appropriate amount, causing a misalignment that decreases the signal the desired amount.

Figure 4-16
Optical power coupler

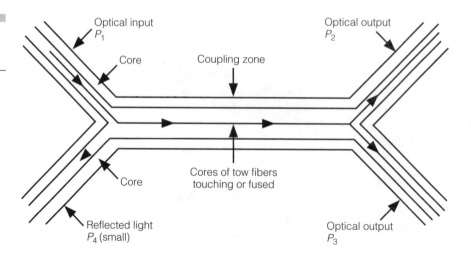

Optical input
P_1

Optical output
P_2

Core

Coupling zone

Core

Cores of tow fibers touching or fused

Reflected light
P_4 (small)

Optical output
P_3

Such splicing gear allows the operator to program in the amount of optical attenuation in dB.

Optical transmitters, receivers, splices, connectors, splitters, couplers, and attenuators are all the devices needed to put together an optical-fiber communications network. Such networks and their terminations can be very large, employing various stations containing a great number of cables and components. Since miniaturization is often desired, optical stations can be mounted in surface cabinets and pedestals or in aerial-mounted housings and closures. Many design choices and methods are available to those charged with designing a system.

4.2 Fiber-Optic Trunking and Cable Television Applications

The first use of optical fiber in cable television applications was to replace the trunk cascade of coaxial cable amplifiers. The replacement benefited cable television systems enormously. Nearly all the system noise and distortion contributions caused by the cascade of amplifiers were eliminated. In addition, eliminating the trunk meant lower power consumption with less related costs and signal leakage from the trunk system.

The only drawback was the fiber-optic signal was essentially light and was not an RF signal. Conversion from the optical signal to an RF signal had to take place at each node or at many nodes. Some progressive urban systems using the subsplit reverse or mid/high-split upstream transmission replaced the downstream (forward) trunk with optical fiber and did away with their reverse systems. As it turned out, few systems made this change, and those that did pulled the forward amplifier modules and used the old trunk system for only the upstream application. The difficulty systems had facing this problem depended in large part on the availability of equipment types their manufacturer had available to rework the reverse-only system.

Systems considering optical fiber additions should study the methods of fiber-optic network topology as well as their cable system network topology. Fiber-optic technology has many similarities to coaxial cable systems. Both cables experience loss, which is a function of frequency, but fiber-optic cables have a difference in loss versus frequency due to the material characteristics of glass. Still, in systems operating at wavelengths of 1310 nm or 1550 nm, the loss is given in dB per kilometer, and the calculation of signal levels is done in much the same way as coaxial

<![CDATA[Chapter]]>

cable. For optical systems, the signal levels are in optical dB/km. The usual unit is in dBm of the optical power level. Handheld optical power meters are used to measure system levels at cable ends.

4.21 Fiber-Optic Cable Overlay

The method of a fiber-optic cable overlay of the coaxial trunk system, shown in Figure 4-17, is the simplest approach in implementing fiber-optic techniques. The figure displays the optical fiber following the trunk route with often two or more fibers serving an area.

Fiber-optic cable is manufactured with many choices of single fibers placed in buffer tubes. For aerial applications, loose-tube cable is often the choice. There may be two to eight fibers per buffer tube and as many as eight buffer tubes per cable. This can result with as many as 64 fibers in a cable. For the fiber trunk method, also known as fiber backbone, 24 total fibers are usually adequate, with plenty of spares (dark fibers) for the future expansion of either up- or downstream requirements.

For cable systems carrying 50 or more channels, some terminal equipment will split the signal band into groups of channels and will optically transmit, for example, 16 television channels per fiber. If the whole channel lineup is transmitted on one laser transmitter on one fiber, the laser

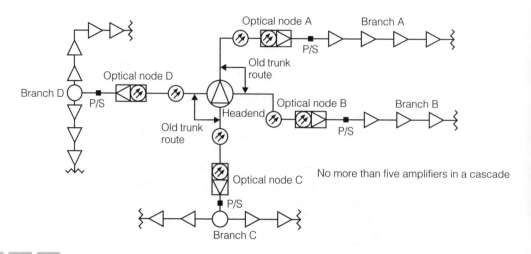

Figure 4-17
Example of a fiber backbone

power is spread over the whole band. When a laser transmitter transmits 16, 6-MHz channels over one fiber, the power is spread over about a 100-MHz band (6 × 16 = 96). This will result in a higher receive power per channel. The network branches, as shown in Figure 4-17, indicate the optical paths that may actually consist of one or more fibers, each carrying a group of television channels. These groups of channels will be recombined at the optical receivers. It should be apparent that the addition of node-connecting fibers will result in redundant optical paths at each node. Such alternate optical signal routing is shown in Figure 4-18. Signal source switching at each optical node can be controlled by a computer at the hub/headend or by a simple loss-of-signal switch.

4.211 The development of system node locations is important in restructuring the network for system upgrades and increasing signal reliability. Alternate fiber-optic routing can result in each node's capability to maintain a reliable, nearly fail-safe system. It is imperative that each node be powered by a standby or uninterruptible power system (UPS). Failure of individual coaxial cable distribution amplifiers will cause only a small number of subscribers to lose service. When the calls come in reporting an outage at a node, the area and the device location often can be determined quickly. Thus, the outage can be corrected in short order. System distribution nodes are another name for a sub-hub, where signals from the main source or headend are supplied by an optical-fiber system.

Figure 4-18
Alternate routes for connecting nodes

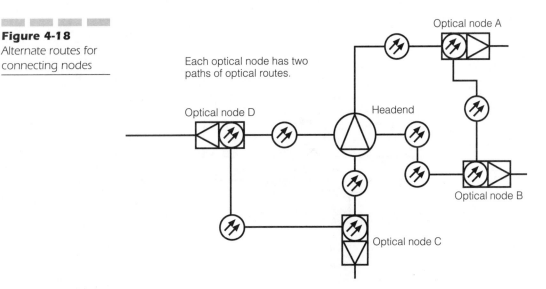

Each optical node has two paths of optical routes.

Optical node A

Optical node D

Headend

Optical node B

Optical node C

The optical-to-electrical (RF) conversion takes place at the node or sub-hub. Signals are then distributed on a normal coaxial cable system to the subscribers' homes.

Fiber to the curb (FTTC) carries this fiber-optic concept still further. The optical signal is converted to RF signals within the pole- or pedestal-mounted tap. The subscriber's home is connected to the tap port via a coaxial drop cable (see Figure 4-19). At present, few if any cable television systems operate in this manner. The day will no doubt come when the fiber-optical signals will be dropped to the subscriber's home and the optical-to-electrical conversion will take place within the set-top modem. This concept is known as fiber to the home (FTTH) and is shown in Figure 4-20.

4.212 Because optical fiber has almost wholly replaced the trunk coaxial cable cascade, some means are needed to supply the return or upstream signal requirements. Instead of using the old trunk system for the return path, a better solution would be to use an optical-fiber system. Enough extra fibers should be included in the cable to cover the return signal needs.

The upstream system is often called the reverse-tree concept; signals that originate from the subscribers are collected along the path to the central hub/headend where they in turn are sent or redistributed to their destinations. Normally, most subscriber signals consist of ordering information for pay-per-view service or possibly utility meter reading information or alarm information. This type of use does not require high band-

Figure 4-19

Optical fiber to RF taps, or fiber to the curb (FTTC)

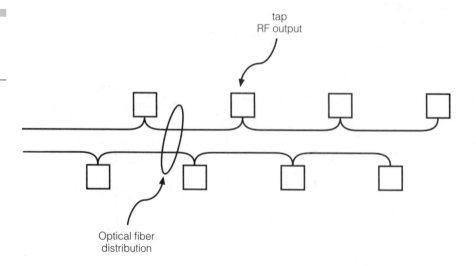

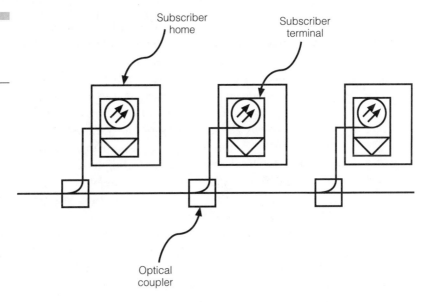

Figure 4-20
Optical signal
distribution to
subscriber terminal

width or high data-rate signals. If telephone service is being contemplated
or computer interconnecting will become a requirement, more upstream
supporting bandwidth will be needed. From a practical standpoint, the
upstream service most likely should be digital. Presently, telephone service as offered by the telco operator functions in the digital domain, beginning with the basic T-1 system. Signals from subscriber locations can be
collected via RF, usually the 5- to 40-MHz sub-band converted to digital
signals at each sub-hub or node, and sent to the main hub/headend on
optical fiber. Each subscriber will have a unique access code contained
in the digital header that will enable the hub/headend to recognize who
is sending or receiving messages. Other header data in digital form will
contain routing and control information.

Several brands of headend and distribution equipment that address
the reverse digital system are presently available. As in the past, some of
the well-known companies have product and application information as
well as customer engineering support available to cable operators considering such plant expansions and upgrades. Also, several consulting and
design companies have suitable software for design and mapping for both
fiber and coax systems. Cable systems using both fiber and coax methods
are known as hybrid fiber coax (HFC) systems.

It is imperative that cable television systems considering staying in the
business a long time add fiber-optic technology to their systems in order
to compete. Many cable systems have merged with telephone systems to

form large conglomerate telecommunications companies. Upgrading such systems with fiber-optic technology and replacing coaxial and copper wire trunk has increased the communication capacity.

The upstream reverse system can be a simple straightforward node to hub/headend or a more complicated node to a sub-hub and then to the main hub point. Many manufacturers and cable operators have and/ or offer several elaborate upstream systems. Figure 4-21 illustrates the node-to-hub system concept. If the return system consists of digitally modulated optical carriers, the digital circuitry at the hub/headend must decode the digital data stream and assemble the data packets to be sent to the desired destination. If the return data is collected on sub-band RF carriers, the whole sub-band can be transmitted on an optical return carrier. This may be simpler than demodulating the return carriers, assembling a high-speed return signal, and transmitting back to the hub/headend as a digital optical signal. Cable systems considering such system enhancements should do a comprehensive survey of available return system components from the various manufacturers.

4.213 Many cable television systems in urban or suburban areas are fortunate to have large apartment or condominium complexes, known

Figure 4-21
Node to hub/
headend concept

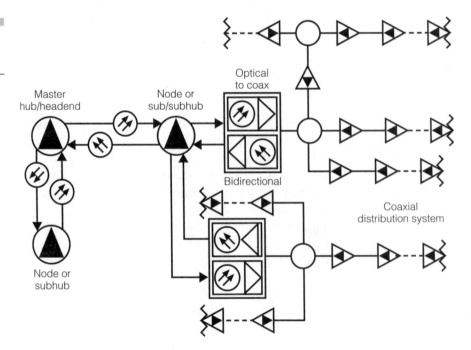

as multiple dwelling units or simply MDUs. These large complexes offer high counts of equivalent dwellings per mile, which is an important criterion for system plant. The internal or inter-building wiring is, in essence, an extension of the distribution system and can add several amplifiers to the cascade. When fiber-optic technology is being introduced into existing coaxial cable systems by replacing a trunk, extending the optical fiber into the complex is advisable. Thus, a node or mini-hub point, fed by optical fiber, should be situated within the building or a group of buildings, depending on the existing coaxial cable network topology.

For high-rise apartment complexes, the fiber node could be located in some location at the middle-floor level. The optical signal can then be converted to RF television carriers and fed to distribution points on the other floors. In some cases, an optical coupler can be connected to a receiver on the first floor and another receiver on the middle floor. The receiver on the first floor can distribute signal from the first to the middle floor and the receiver at the middle can distribute signals to the top floor. Thus, many design choices are available with various trade-offs. For each situation, the best choice is often the obvious one. Figures 4-22 and 4-23 illustrate a couple of design examples.

Figure 4-22
Optical to RF coaxial on first floor

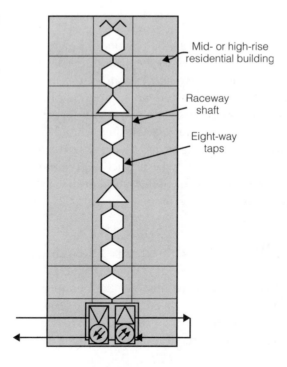

Mid- or high-rise residential building

Raceway shaft

Eight-way taps

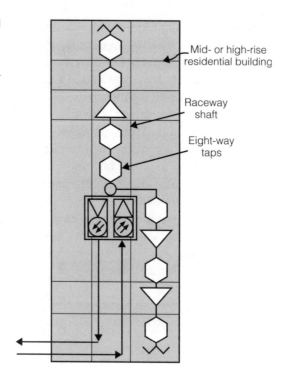

Figure 4-23
Optical to RF coaxial
mid-level

Mid- or high-rise
residential building

Raceway
shaft

Eight-way
taps

Because many apartment or condominium complexes have professional people as residents, a high-speed computer data service will be welcome. This means that the two-way return system has to be active. Therefore, the fiber-optic return system should be installed to assemble subscriber return signals and connect them to the hub/headend for further routing. The downstream system with its large bandwidth will send the digital information back to the requesting subscribers. This high-speed Internet connection will surely be attractive to many subscribers. Remember that the high-bandwidth cable system drop is far greater than the copper twisted pair of the phone drop.

4.22 Fiber-Optic Super-Trunking

Optical fiber systems are ideal to use in super-trunk situations. Super-trunking methods have been used as a good solution to a variety of problems. Such applications consist of satellite receiving connections to the hub/headend point or connecting hubs together. Connecting hubs together improved the signal reliability but added significant power and maintenance costs. Also, more coaxial plants mean more possible points for RF leakage and ingress problems. Replacing super-trunk systems with an optical-fiber system provides all the good points and few of the bad points.

4.221 The satellite to hub/headend connection is the most common use for a super-trunk. Realize that a super-trunk system is just a signal transportation method from one point to another, usually in one direction. Only one fiber is typically required to carry signals from a satellite station to a hub. But, most manufacturers' cables contain a number of fibers in color-coded buffer tubes. When optical fiber is planned to connect the downstream signal and upstream return path from the mini-hubs or nodes, a single fiber or more can be used to pick up the signal from the satellite-receive station to the master hub/headend. This concept is shown in Figure 4-24.

Systems using a coaxial cable super-trunk connecting a satellite station to a hub/headend a distance of 5 mi could use an amplifier cascade of two or three amplifiers, depending on the cable size. Such a cascade should have its own power supply so that it is not dependent on the downstream path for its power. The frequency plan for transmission from the satellite station can be lowered to the standard VHF band for the upstream run to the hub/headend in order to be in the lower-loss end of the coaxial cable.

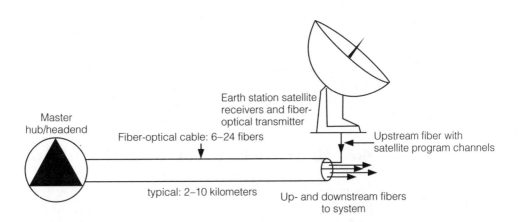

Figure 4-24
Fiber-optic Earth receive station to master hub/headend system

At the hub/headend, the satellite channels have to be converted for the required downstream channels, yet all this conversion adds noise and distortion to the signal. When fiber-optic methodology is used, the "on-channel" assignments can be made at the satellite station and transmitted to the hub/headend. The optical receiver will supply the "on-channel" RF signal to the combining network for downstream distribution. From this example, the benefits of using optical-fiber technology should be evident.

4.222 As cable operators start to utilize fiber-optic technology, more and more uses become evident. When installing fiber-optic cable to replace the coaxial trunk and expand the upstream return bandwidth, extra fibers are often included in the cable for future use. In many instances, these extra fibers (dark fibers) may not be seen as necessary at the time of installation, but their use will become evident later on.

One such use is to connect to the local broadcast television stations by optical fiber. Recall that early cable systems' main reason for operating was to improve the reception of broadcast television stations. Using optical fiber, television broadcasts can be received from distances up to 50 km from the stations. By doing so, the quality of the programming will certainly be high. Gone are the industrial sparklies, weak signal snow, and lightning interference. Even if the station's transmitter fails, the cable signal should still be there. Employment of the emergency broadcast service (EBS) would also be enhanced because all stations are required to carry this service and all cable systems are required to participate.

The system design for broadcast station pickup is nearly the same as for the satellite-receive station pickup, shown in Figure 4-24. Variations can be configured, depending on the location of the studio or transmission facility. Preferably, the station would have available a video and audio feed point so the cable operator would have a choice of transmitting base-band video and audio or use an on-channel television modulator transmitting optically to the hub/headend location.

4.23 Benefits of Fiber-Optic Plant Addition

There are indeed great benefits to cable television operators who install fiber-optic systems, including cost savings, signal quality improvements, and signal reliability. Cable television operators have been striving for years to improve the product and control costs, both of which are attainable by adding optical-fiber technology. In this section, we'll take a brief look at the factors that are involved in adding fiber optics to cable systems.

4.231 Most cable operators have problems controlling costs when making plant improvements have to be balanced between the costs and the prospective earnings and/or savings. Because one single-mode fiber in a multifiber cable has the capacity to carry more information bandwidth than a coaxial cable, a cost comparison is very difficult. Also, a single fiber connecting services over normal cable system distances requires no repeater amplifiers or electronics between the transmitter and receiver. Thus, comparing the amplifier coaxial cable cost per mile with a single fiber might be a more meaningful comparison. However, in aerial and underground systems, a normal fiber-optic cable contains several fibers placed in several buffer tubes. Hence, a single- or dual-fiber cable is not normally manufactured for aerial or underground plant.

For those more mathematically inclined, a more analytical approach would consist of a calculation of cost per hertz of the bandwidth. Such a procedure might be useful in convincing management to invest in an optical-fiber system upgrade. A rudimentary cost analysis for a run of cable may give some comparison between coaxial and fiber optics. Excluding labor costs for strand and cable placement is a reasonable consideration because it is assumed to be the same for coaxial and optical plant. Example 4-1 shows the cost factors and calculations for 750-MHz cable service for a 5-mi sample run. Remember that the design topologies between the two methods can become quite different.

4.232 One of the benefits of fiber-optic systems is the improved signal quality. The technical staff is well aware of the noise buildup and distortions of a cascade of amplifiers. As the upper frequency limit of coaxial cable systems increases, the number of amplifiers per mile increases. This fact will in turn limit the length of a cable run or system reach. Some operating cable systems that employ long cascades of costly and power-hungry feed-forward amplifiers may find it is impossible to extend the bandwidth of such systems without doing a fiber-optic rebuild.

Some system operators use a microwave feed from a master headend facility and beam programming to several sub-hubs serving the areas with cable service. Each sub-hub has a microwave receiver and electronic equipment that converts the microwave signal to the normal cable television channels. The master headend may have several transmitters and antennas to beam the signal in several directions. Such a system avoids

Example 4-1

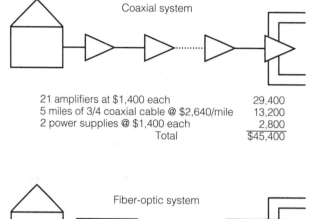

Coaxial system

21 amplifiers at $1,400 each	29,400
5 miles of 3/4 coaxial cable @ $2,640/mile	13,200
2 power supplies @ $1,400 each	2,800
Total	$45,400

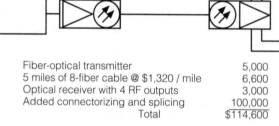

Fiber-optic system

Fiber-optical transmitter	5,000
5 miles of 8-fiber cable @ $1,320 / mile	6,600
Optical receiver with 4 RF outputs	3,000
Added connectorizing and splicing	100,000
Total	$114,600

the amplifier cascade distances and provides improved signals to the sub-hub distribution points. Maintenance of such systems, along with following the FCC rules of compliance for the facilities, is usually costly. Some cable systems have replaced the microwave facilities with optical fiber connecting the sub-hubs to the master headend.

4.233 One of the most important benefits for adding optical fiber in place of coaxial system trunk is that system reliability is improved. As stated earlier, often no electronic devices are used in fiber-optic runs for distance up to 30 miles. Thus, there is nothing that can fail or cause problems unless it is subject to a disaster such as a fire, or having a pole hit or the cable cut.

Some cable operators serve a node or sub-hub by dual optical paths, thus adding a redundant feed to each distribution point. If problems do occur, it usually is the coaxial plant that causes the outage. Optical fiber runs do not contain connectors that can pull out of amplifiers or line passive devices that cause outages. Also, RF leakage or ingress does not happen in fiber-optical systems. Therefore, signal quality and reliability are maintained to superior standards.

4.3 Fiber-Optic Construction and Installations

Fiber-optic cable installation methods have a lot in common with coaxial cable installation, but they have differences as well. One must remember that the information-carrying optical fiber is glass, which is brittle and sharp and can shatter, while coaxial cable is metallic. Both types of cable employ plastic jackets for protection from water and handling. Coaxial cable has mechanical strength in its center conductor, outer aluminum sheath, and its plastic jacket. Fiber-optical cable has a metallic or tough plastic strength member and outer jacket that adds strength. As mentioned earlier, the optical fibers are either tightly or loosely contained in color-coded buffer tubes and are appropriately named loose-tube or tight-tube construction. In most cases, loose-tube optical cable is used in aerial plants, while tight-tube cable is often used in interbuilding or underground plant applications.

4.31 Handling Fiber-Optic Cable

Handling optical cable carefully is important in order to avoid damage and ensure that the cable will have a long service life. Optical cable should not be bent beyond the minimum specified bending radius. It should also not be subjected to heavy objects being placed on it or running over it. Such action can stress or damage the glass fibers within the cable. Also, the cable should not be subjected to physical shocks either from hitting the cable or dropping a reel on the pavement. Most of the damage to a fiber-optic cable occurs during the loading and unloading delivery phase or during installation.

4.311 The initial loading of cable is done at the manufacturing plant and is usually done correctly using proven methods and equipment. If the shipment to a cable operator is direct from the manufacturer on the manufacturer's truck, minimal problems can be expected. Fiber-optic cable that is shipped to a distributor and then to the end user usually has to go through two or more load and unload cycles, where more possible damage to the cable may result. Therefore the cable system operator should be aware of where and how the cable is being shipped.

Most cable manufacturers and sales materials list instructions for loading/unloading procedures in their catalogues. Generally, all manufacturers of optical-fiber cable state that cable reels should stand on the edge of the reel and should be rolled in a specified direction. The reels themselves are made of either steel or heavy wood. A steel reel can usually be returned to the manufacturer for credit, while the wooden reels may not be returnable. The wooden reels are fully lagged and strapped with two steel bands around the lagging. The lagging consists of wooden slats nailed to wooden reel flanges. This protects the cable from being smashed, as shown in Figure 4-25.

Unloading from a trailer truck can be done with a ramp, but it can be tricky because the reel has to be restrained from rolling on the ramp. Probably the best way is with a crane or a forklift truck. It works well to place a pipe through the reel hole and use a crane or forklift to lower it to the ground, one reel at a time. Usually a reel of optical cable contains more footage than coaxial cable; hence, the number of splices is held to a minimum. Some manufacturers' trucks are equipped with either a crane or lift-tailgate, making the unloading procedure simpler.

4.312 Proper storing of optical cable is important and necessary while the optical system is being installed. Fiber-optic cable should be left standing on the reel edge or rolling edge and should never be stacked reel-on-reel on its side. It should be separated from coaxial cable so it is not mixed up by the installation contractors and handled unnecessarily.

Indoor storage is best for optical cable since wet and cold weather can cause deterioration of the wooden reels. Systems storing optical fiber in a warehouse should use either an appropriate loading dock or a forklift truck to place the optical cable reels on a reel trailer for installation.

Figure 4-25
Example of a
properly made reel
of optical fiber cable

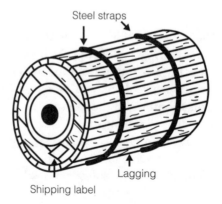

Steel straps

Lagging

Shipping label

4.32 Admittance Testing

As with most commercial companies, testing incoming material should be performed and any equipment or material ordered that does not meet the necessary specifications can be returned. This process prevents installing faulty equipment that produces a poor system or product. Admittance testing of materials should be performed on all equipment, not just cable. The testing of optical cable for loss (attenuation) and effective length can be done quite easily on a reel-to-reel basis because cable manufacturers make both ends of the cable accessible on a reel.

4.321 Cable attenuation can be tested simply with a low-cost power meter and light source. Some sources and meters can have the bare ends of the fibers cleaned and cleaved. Since this process is done on each fiber, placing connectors on each one is a big job. Some cable operators test one or two fibers in a cable and assume if they test well, then the rest must be okay. It is best to test each fiber individually. This procedure is shown in Figure 4-26. Both ends of the cable have to be accessible for this method.

Figure 4-26

Testing for cable attenuation

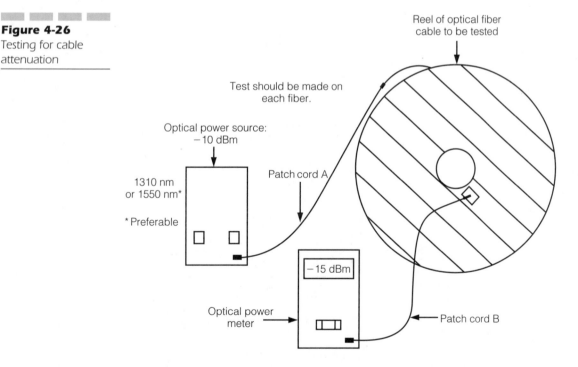

Reel of optical fiber cable to be tested

Test should be made on each fiber.

Optical power source: −10 dBm

1310 nm or 1550 nm*

*Preferable

Patch cord A

−15 dBm

Optical power meter

Patch cord B

A better method of testing for optical attenuation is to use an optical time domain reflectometer (OTDR). This method shows any abnormalities in attenuation for the whole length of the optical fiber. Since OTDR instruments cannot make measurements close to the instrument cable connector, a length of a single fiber is placed between the instrument and the reel of fiber-optic cable. Essentially, this moves the cable reel to be tested away from the instrument by the length of the single fiber, which is usually approximately 1 km in length. This blind spot of the OTDR is referred to as the dead zone. This single-fiber cable is tested and the results are documented before testing the cable reels. This procedure is shown in Figure 4-27.

4.322 Once the testing of incoming fiber-optic cable has been completed, the results should be recorded and filed in an appropriate manner. This

Figure 4-27
OTDR testing of
fiber-optic cable

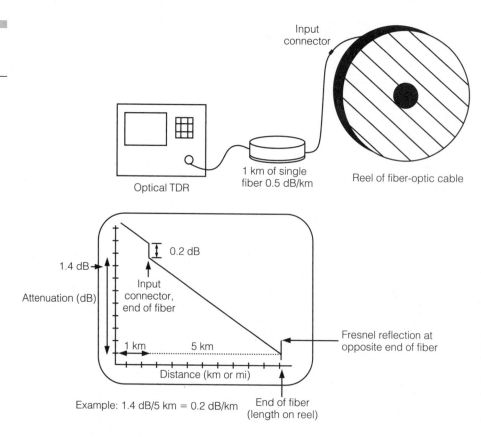

Input
connector

Optical TDR

1 km of single
fiber 0.5 dB/km

Reel of fiber-optic cable

0.2 dB

1.4 dB

Input
connector,
end of fiber

Attenuation (dB)

Fresnel reflection at
opposite end of fiber

1 km   5 km

Distance (km or mi)

Example: 1.4 dB/5 km = 0.2 dB/km

End of fiber
(length on reel)

information could prove useful if a problem occurs after installation. The documentation can simply be a list of each fiber and its attenuation where the fiber is identified by the color of its outer coating and the color of the buffer tube containing it. Each reel also has a serial number that should be recorded. Some cable operators develop a printed form made into pads so they can be numbered, torn off, and filed. An example of such a form is shown in Figure 4-28.

A faster and more elegant method is to have an OTDR that has a digital memory and readout feature. Such instruments can store measured data in digital memory for later examination. Fiber color and buffer tube color identification information can be entered via a keyboard that comes with the instrument. This process allows the operator to enter the reel number and measured attenuation, along with other information such as the instrument, operator, time, and date. Data storage can be done on a convenient floppy disc.

Figure 4-28

Example of a manual record of test data for fiber-optic cable

Company Name, Address, and Telephone Number

Test date:_____   Name of technician:_____

Cable reel no:_____   Order no:_____   Date of delivery:_____

Type of installation:_____   Inst. model no:_____   Calib. date:_____

Test wavelength:_____ nm

Admittance Test Data

Buffer tube color	Fiber color	Optical att. (dB)	Buffer tube color	Fiber color	Optical att. (dB)

4.33 Fiber-Optic Cable Installation

Installing fiber-optic cable requires some special procedures that are different than the procedures used for installing solid aluminum-sheath coaxial cable. Most people remember quite well that coaxial cable should not be crushed or kinked during handling and that the installation process can cause the cable to be stressed more than the loading and unloading process. Therefore, the differences in handling will be addressed for both the aerial and underground installation process.

4.331 In general, fiber-optic cable is manufactured in longer lengths than coaxial cable for two reasons. First, it is not necessary to place amplifying or repeating devices along the cable run, and second, it is most important to minimize the number of splices. Splicing fiber-optic cable can be an extremely long and laborious process because each fiber has to be individually fusion spliced and placed in a splice tray. The splice trays are then placed in cylindrical PVC weather- and moistureproof enclosures. The long lengths of fiber-optic cable have to be installed in sections. Therefore, pulling tension on the cable has to be closely monitored in order to avoid exceeding the recommended value by the manufacturers. A process called figure-eighting is often used, which helps relieve pulling tension. This process is shown in Figure 4-29.

Contractors who are more experienced and specialize in optical-fiber cable installation have equipment that controls the tension on the cable.

Figure 4-29
Figure-eight tension relief method

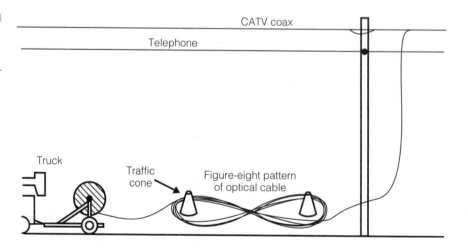

Such equipment is a servo-controlled winch or capstan drive where the cable tension controls the action. Pulling grips used for optical cable are the same type as used for coaxial cable and distribute the tensile pull load over the outer cable jacket.

Because splicing should be avoided, service loops (excess cable) are placed along the cable run. The initial loop is placed at the beginning of the run and often appears approximately every fifth or sixth pole span. This excess cable can be used when the aerial pole plant is rerouted or when new roads or road-widening work is needed. With this advance planning, cutting and splicing in a new piece is avoided. The excess length has to be designed into the system and initially taken into account when calculating cable attenuation and optical input/output levels. These service loops use a turning frame that resembles a snowshoe or tennis racquet. Essentially, a figure-eight pattern of cable is stored in the loop and is raised and mounted on the steel aerial strand. An example of a service loop is shown in Figure 4-30.

In general, fiber-optic cable should not be tightly lashed with coaxial cable, so precautions should be taken when pulling the lasher. It is vital to find a contractor who is experienced in proper fiber-optic cable installation.

Figure 4-30
Excess cable stored
in service loop

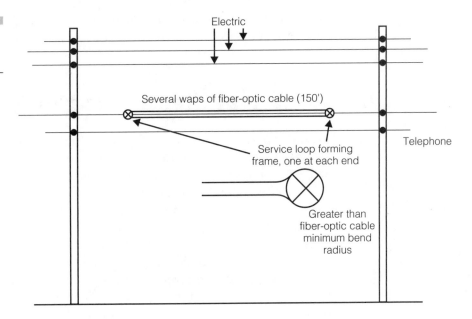

4.332 Underground cable installations seem to be used more often in today's urban and suburban environments. Underground cable installations use a variety of methods and techniques from direct burial in a trench to a conduit system. Conduit appears in a rigid pipe form of PVC or semiflexible corrugated plastic tubing. Some conduit manufacturers can supply the cable preinstalled in the flexible conduit, which can then be direct buried in a premade trench or plowed directly.

In city or urban areas, optical cable is pulled through existing ducts. In such installations, a pulling tape is placed in the duct by a compressed-air method. Next, the fiber-optic cable is connected to a cable grip, which in turn is fastened to the pulling tape. A winch at the other end pulls the tape through the duct, followed by the fiber-optic cable. A pulling lubricant that aids in relieving the pulling tension is applied at the source end of the cable run. The pulling tension should be monitored at the pulling end by a dynamometer or by a servo control on the winch. The maximum pulling tension should not exceed the value recommended by the manufacturer. The splice locations for the underground plant should be placed conveniently for servicing. Some of the equipment used was shown earlier in Figure 1-9 in Chapter 1.

Experienced contractors should be hired to install underground optical cable plant. In areas where trenched or plowed cable is installed, it is a good idea to install a metallic conductor along with the optical cable in order to facilitate identifying the cable path with a locating instrument. If a coaxial cable is to be installed along with the fiber-optic cable, the metallic sheath of the coaxial cable can be used to place the locating signal.

4.4 Aerial Electronic Equipment

As with any cable system, certain electronic devices have to be placed along the way. Such electronic equipment requires some power source, and it certainly cannot get it from a glass fiber. Either the fiber-optic cable has to contain metallic conductors to carry power or it must be obtained from the coaxial cable plant. Since optical-fiber cable has replaced the coaxial cable trunk system, any power needs are taken from the coaxial cable distribution system. Such equipment consists of the receivers that convert the optical signal to the RF signal for distribution to subscribers. At these locations, the reverse signals are assembled and are converted to digital signals or transmitted in the analog domain in the sub-band through an upstream transmitter using upstream optical fibers.

4.41 Types of Electronic Equipment

At locations where the fiber-optic cable feeds signals to the coaxial distribution system, the optical signal may terminate or be passed through to a location further downstream. An optical coupler can be used to divide the signal to feed the optical receiver with an appropriate optical level and pass a larger level on farther. This is essentially the same method used in the RF directional coupler and is described in Figure 4-31. An optical transmitter can be incorporated in the terminal package to retransmit the optical signal farther downstream.

4.411 Several manufacturers make optical couplers that essentially allow a tap leg and a through leg. Such devices are usually fusion spliced into the system and are placed within the housing containing the optical receiver. Since most aerial housings are two pieces with a hinged cover and weather-sealed gasket, as well as RFI gasket, the optical portion usually occupies one side and the RF the opposite side. This is similar to housings used in coaxial cable systems with the usual RF test point available for signal-level measurement. The advantage of placing pigtails and connectors to the output ports of the optical coupler is that it permits measurement of the optical signal using a power meter. In many cases, the optical signal provides service on several fibers connecting to several distribution nodes, as shown in Figure 4-32. Nodes can be sub-headends or hubs where locally generated programming can be inserted. Some systems distribute the signals unscrambled to the hub sites where local signals are inserted and premium channels scrambled.

4.412 The optical receiver placed in a node or sub-hub location converts the optical signal to an RF signal consisting of all the television carrier frequencies. A nominal RF output level is usually between +30 and +35 dBmV, with most manufacturers at +32 dBmV. Often the maximum and

Figure 4-31

Split downstream signal from optical coupler, upstream on a separate fiber

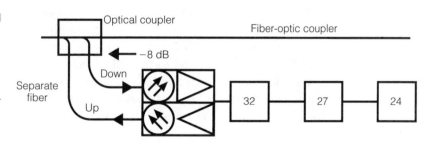

Figure 4-32
Nodes being fed
cable service on
separate fibers. (Only
the forward system is
shown.)

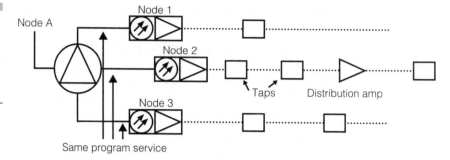

minimum optical signal levels that provide a range of RF output signal levels are specified. Equipment manufacturers usually have a variety of options available to the system designers. Such options may be a single, dual, or quadruple RF output. Thus, the optical receiver system can act as an RF bridging amplifier, providing signal level up to four outputs. Another feature is the return option that assembles the upstream RF carriers and converts them to an optical signal for transmission to the head-end hub via an optical fiber. If the upstream signal operates on a different wavelength than the downstream signal, the same fiber can be used to carry both up and down signals. In this case, a splitter-coupler combination has to be placed in the fiber at the headend/hub and at the node location. Such a system is illustrated in Figure 4-33.

4.413 The optical equipment placed at node or sub-hub locations usually receives electrical power from the coaxial cable RF plant. This power source is the usual 60- or 90-v.a.c. power from the normal cable plant. Because this optical receiver is critical for providing service to the node distribution point, a standby or UPS power supply is necessary.

Remember that system reliability is a requirement for any system operating in today's telecommunications environment. In many instances, the sub-hub or node is not a strand-mounted or pedestal-mounted unit; rather, it appears in a building housing local signals. Dropping services and adding services are performed at these locations, and because commercial power is available, the optical receivers are rack-mounted where the RF channels are used to provide selected programming, as required by the sub-hub. The sub-hub location should have a standby power source to provide emergency power. Often at these locations RF television signals provide service to the immediate area, while remote service nodes are served by optical fiber.

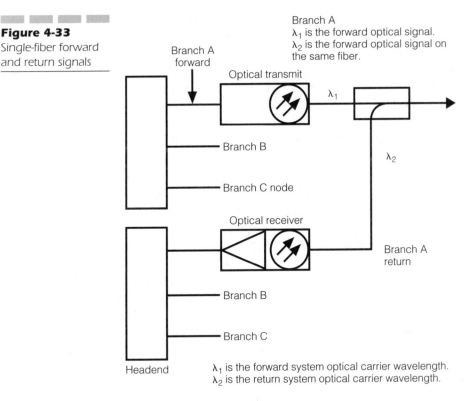

Figure 4-33

Single-fiber forward and return signals

Branch A
λ_1 is the forward optical signal.
λ_2 is the forward optical signal on the same fiber.

Branch A forward

Optical transmit

λ_1

Branch B

λ_2

Branch C node

Optical receiver

Branch A return

Branch B

Branch C

Headend

λ_1 is the forward system optical carrier wavelength.
λ_2 is the return system optical carrier wavelength.

An early approach to emergency powering of remote electronic equipment was to use a rechargeable battery pack built into each device. This plan did not work out because most available battery packs did not provide enough "on" time to keep the device operating during prolonged outages. The day may arrive when battery packs are improved and power requirements are reduced, making this method practical.

All service vehicles should have a substitute commercial power source installed. This makes it possible for vehicles to provide commercial substitute power to the standby power supplies when the batteries are exhausted.

4.42 Optical Equipment Tests

Remote optical equipment appears either as strand- or pedestal-mounted devices servicing sub-hubs or mini-hubs. Here the conversion from optical to RF television carriers takes place. Some systems house their terminal equipment in buildings where commercial power is available. Such equipment is usually rack-mounted. At these locations, optical and RF signal levels can be deleted or added.

4.421 Measurement of optical signals at the receive sites is necessary to establish the optical signal input to the receiver. When setting up a system, the transmitter at the hub/headend is turned on and the optical signal is measured at its output. When the optical fiber is connected to the transmitter, the optical signal level is measured at the receive point. The difference in levels is the cable loss. These levels are measured with an optical power meter indicating the level in dBm of optical power. The difference in dBm readings from transmitter to receiver is the optical cable loss in dB.

Next, the signals are connected to the transmitter and the power level in dBm at the receiver is recorded. This value can be written on a label at the receiver for future reference. Now the RF signals can be measured and adjusted, as required by the coaxial cable system. Such adjustments are for level and slope of the television carriers. Most optical power meters are battery-powered and operate in either a single- or dual wavelength mode. Such optical power meters have a variety of features, such as the input-connector type, a set of adapters, extra battery packs, as well as a carrying case. Some power meters have an accompanying light source available to aid in making fiber-optic cable-loss tests. As stated earlier, readout is in optical power level in dBm (1 mw reference). These instruments cost anywhere from $500 to $1,500, depending on model and options.

4.422 After the optical fiber system has been installed by the subcontractor and the actual lengths of optical cable has been substantiated, a rigorous testing program should be conducted. Essentially this follows the same information from the design phase of the project. All the losses in each section are measured and compared to the projected values assumed in the design phase. Such losses include optical-fiber loss, connector loss, and splice loss, plus an assumed margin loss (usually 2 or 3 dB). If the measurement of these losses confirms that the assumed losses used in the design are close, it can be assumed the optical-fiber system will perform according to specifications. See Table 4-2 for an example of optical loss measurements.

The measured data should be measured at the design optical wavelength using a light source and optical power meter. Reasonably priced, accurate optical time domain reflectometers (OTDR) are available for measuring total link loss. These instruments can make the measurements on most

Table 4-2

Losses for an
Optical Cable
System[a]

Component	Specific Loss	Measured Loss (in dB)
Optical fiber	(1.0 dB/km) 20 dB	16
Splice loss	(0.2 dB ea) 4 splices, 0.8 dB	1.0
Design margin	2 dB	2
Total Loss	22.8 dB	19[b]

[a]For 20 km of cable at 1300 nm

[b]Notice that the margin is 22.8 − 19 = 3.8 dB better than the specified loss.

of the usual optical operating wavelengths. The measured loss includes the complete amount of cable installed including the lengths stored in the service loops (estimated at 0.1 km/loop).

When the transmitter is ready for installation its power output can be measured using the portable power meter. This value should be compared against the manufacturer's specification for accuracy and can be used as the input power to the optical fiber when connected. At the receiving end, an optical power meter can now measure the power level that will be connected to the receiver. The value of this power level should be within the dynamic range of the receiver as specified by the manufacturer. Example 4-2 provides further discussion.

Example 4-2

If an optical transmitter with an output power of −5 dBm connected to the fiber-optic cable system with a measured loss of 19 dB, the receiver should have an input power level of −24 dBm. If the receiver has a dynamic range of −12 to −30 dBm, then −24 dBm power level is within this range with room to spare.

4.423 Once the proper optical levels have been established and compared to the predicted design parameters, the RF system has to be adjusted. These measurements can be made using a signal-level meter. For this method, measurements are made on a channel-by-channel basis. Some signal-level meters have a wideband-tuning feature that presents all the carrier levels on a wide LCD similar to a spectrum analyzer. The method of choice would be to use a high-quality spectrum analyzer, preferably one with digital storage. Measurements can be made and recorded on a printer either on location or later at the office. These printouts are useful maintenance tools for monitoring system performance. Long-term data can be stored electronically for comparing signal levels over time. A high-quality spectrum analyzer, as used by many cable systems, can measure signal-to-noise ratios as well as signal distortions. Cable operators using fiber-optic technology usually can see the difference in signal quality at the node sub-hub locations as compared to the cascade of amplifiers previously connected to these points.

Summary

The development of optical-fiber systems as a method of signal transmission and reception provided a much needed solution for cable systems faced with extending their plant beyond the coaxial cable system capabilities. Coaxial cable systems are limited by the cable loss and the number of RF electronic amplifiers in cascade. Essentially the signals from the headend to the end of the system are distance limited.

A study of the loss of optical fiber and the bandwidth capabilities of transmitter-receiver combination exceeded that of a coaxial system. The loss characteristics of optical fiber are enormously better than coaxial cable, as well as the bandwidth. The same distance covered by an optical fiber, in many cases, has no electronics between the transmitter and receiver, whereas if the cable run were coaxial a cascade of power-hungry amplifiers would be required. At the same time, it's clear that optical fiber also has problems. It is difficult to manufacture, is relatively costly, has to be handled gently, and is expensive and hard to splice and connectorize. For cable television systems, optical fiber cable is used to replace the coaxial cable trunk system. At nodes, the signal is converted from the optical domain to the RF television domain. Gone are the long, high-maintenance and power-hungry trunk amplifier cascades that accumulate noise and distortion products that degrade the television picture quality. Cable systems employ optical fiber to lower maintenance costs and signal leakage and to improve signal quality. Fiber-optical technology is used not only in the trunk system but in connecting the television broadcaster's studio for direct signal pickup instead of using a tower-mounted off-air method.

An optical-fiber connection from the Earth station to the hub/headend is another application. Optical fiber systems have replaced some microwave links. As cable television systems expand their cable plant to provide more services, optical-fiber systems are going to be the method of choice.

Questions

1. Explain how a light-emitting diode (LED) can be used as an optical transmitter.

2. Give two main reasons a solid-state laser is better than an LED optical transmitter.

3. Why is single-mode optical fiber the best choice for cable television system applications?

4. Explain why a roll of optical-fiber cable has to be handled delicately.

5. Describe the operation of an optical time domain reflectometer (OTDR).

6. Explain why optical fiber is not used in the distribution network of a cable television system.

7. What electronic device is required at the optical-to-coaxial cable interface?

8. Power is injected into coaxial cable systems to power the various amplifying equipment. Because optical fiber is glass (an insulator), how is power distributed?

9. Explain the use of a light source and an optical power meter in testing a section of optical fiber.

10. Describe the difficulty in splitting or signal dividing in optical fibers.

Problems

1. Compare the loss of an optical fiber with a loss of 0.32 dB/km to a length of coaxial cable with a loss of 1.3 dB/100 ft.

2. If an optical fiber is going to be connected by a connector-to-connector patch panel arrangement, (a) calculate patch panel through loss if each connector has a loss of 1 dB, and (b) compare loss against a fusion splice with a loss of 0.1 dB.

3. A reel of optical cable is labeled as having a length of 2 km. If the specification for an optical fiber states a loss of 0.5 dB/km, what must the power meter read if an optical source injects a -10 dBm optical signal into the opposite end if the length as labeled is correct?

4. If the manufacturer's specification states a loss of 0.8 dB/km, calculate the loss for a 15-mi length of optical cable.

5. Continuing with the 15-mi optical-fiber run in problem 4, if there are splices at every 2 km of cable length, what is the accumulated loss if each splice has a loss of 0.05 dB?

6. If a -23 dBm signal level is required at the receiving end of the optical cable in problems 4 and 5, how much power is required of the transmitter?

Digital Technology and Cable System Applications

Objectives

After learning the material in this chapter, the student will be able to

- Understand the differences between analog and digital data and the related merits of each.
- Describe the methods of digital data storage and processing.
- Explain what is known as data communication and data processing.
- Describe data-communications systems such as local area networks (LANs) and the various LAN topologies.
- Explain the methods of using the telephone system network for digital data communications.
- Describe the theory and practice of applying optical fiber as a data-transmission medium.
- Explain digital video methods and the application to digital television transmission and high-definition television (HDTV).
- Describe the present standards for cable and broadcast television digital signals.
- Explain the added digital services, such as video-on-demand (VOD) and voice over the Internet protocol (VOIP), for a cable communication system to offer its subscribers.

5.1 A Short History of Digital Communications

Many of us feel that it is becoming a digital world with all the computers, cellular telephones, and digital television. Actually, digital technology has been with us a long time. Digital technology involves the binary numbering system with two digits, or two states such as *on* and *off*. Connecting and disconnecting circuits is the business of the telephone companies and it should come as no surprise that much of the development of digital technology came from the telephone companies, mainly the well-known Bell Labs.

Two-state communications was used by primitive tribes in the forms of smoke puffs or drum beats. Morse code is a form of digital code; long tones and short tones are used in combination to determine alphanumeric characters. A human operator either writes down the message or types it on a typewriter. Ships at sea used Morse code transmission and receivers to communicate with shore stations and other ships. Long-distance communications were realized using such techniques quite reliably. Marine radio used Morse code for ship safety until recently. Morse code marine radio has been terminated and ships at sea no longer require a qualified operator on board. Now, high-speed satellite communications systems operating with digitally coded message packets provide marine communications.

5.11 Nature of Digital Technology

Because digital systems operate in a two-state manner, digital communications are similar to a conversation in which one can ask only "yes" or "no" questions. Because two states can be described by two voltage levels, such as 0 volts and +1 volt corresponding to a binary 0 and binary 1, then a train of pulses can describe a sequence of binary numbers. It is these pulse sequences that describe digitally encoded alphanumeric characters. A pulse train of 0-volt (binary zero) and +1-volt (binary one) pulse sequences transmitted through a wire will have a d.c. level that is positive and varies with time. Therefore, reworking the digital pulse stream to a better form will save transmitting power levels through a wire or cable. Two types of digital sources communicating through a communications channel are shown in Figure 5-1. In Figure 5-1a the communications channel is digital and in Figure 5-1b, the communications channel is analog. These types are essentially the methods used for computer communications through the telephone system.

5.111 As mentioned earlier, digital systems are based on a two-symbol binary numbering system in which sequences of pulses of voltage levels correspond to binary digits. In this system, there is a direct relationship between a decimal magnitude and a binary magnitude. Thus, one can convert from one to the other, as shown in Figure 5-2. It is this relationship between the decimal and binary numbering systems that allows the analog value to be converted to a digital (binary) value. Analog-to-digital converters basically take an analog time-varying electrical signal, sample it, and convert each slice or sample to a digital number representing the amplitude of the sample. Figure 5-3 describes this process.

Figure 5-1

*Digital
communication
examples*

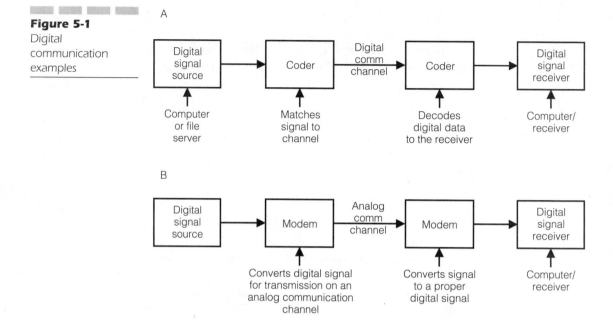

A

Digital signal source → Coder → Digital comm channel → Coder → Digital signal receiver

Computer
or file
server

Matches
signal to
channel

Decodes
digital data
to the receiver

Computer/
receiver

B

Digital signal source → Modem → Analog comm channel → Modem → Digital signal receiver

Converts digital signal
for transmission on an
analog communication
channel

Converts signal
to a proper
digital signal

Computer/
receiver

Figure 5-2

*Decimal to binary
conversion*

For the decimal quantity of 24,
the digital number corresponding
to 24 is computed as follows using
the scale shown below:

Least significant bit

Most significant bit→ 7   6   5   4   3 2 1 0   Bit number

2^7 2^6 2^5 2^4 2^3 2^2 2^1 2^0 Value

128 64 32 16 8 4 2 1

If all 8 bits (0–7) are used to describe
decimal 24:

16 + 8 = 24 = 0001100

It should be noted that each sample value is held from the sample signal to the next sample. During this time, the ramp signal is still increasing, resulting in an error in voltage value between samples. Increasing the sampling rate, thus decreasing sample time, will result in a smaller error. Notice there is a 4-bit digital number representing each sample; doubling the sample rate will double the number of samples and the samples per second.

Figure 5-3
A voltage ramp signal
in analog and digital
form

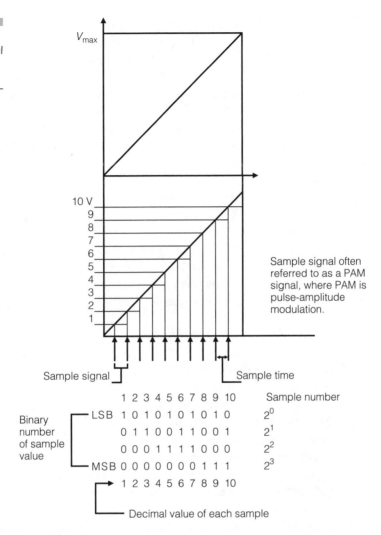

Converting analog voltage values to digital numbers is a common technique used in digital voltmeters and multimeters. The digital numbers are then displayed as decimal digital numbers. A chip set is used in the digital multimeters.

5.112 Now that it has been shown that electrical values of a voltage or current can be represented in the digital domain as a sequence of digital numbers, many uses and methods can be developed. Using the example of

Figure 5-3, the sequence of the 10 samples can be represented, as shown in Figure 5-4. This digital pulsed signal has a +1 volt as a binary one and a 0 volt value as a binary zero. This signal is often referred to as unipolar (either 0 or 1 V) and so has a varying d.c. value, depending on the amount of binary ones (positive pulses). This can become a problem when transmitting digital information through a communication system.

Another technique that reduces the d.c. level is to use a +1 V for a binary one and a −1 V for a binary zero. This method tends to reduce the d.c. level value and is often referred to as bipolar. These pulse trains can also split the signal time of a bit and return to a zero value before changing to the negative value representing a binary zero. This type of signal is referred to as RZ (return to zero). Several pulsed signals representing the same binary number is shown in Figure 5-5. Circuits using such pulsed signal waveforms maintaining bit synchronization are most important. Clearly, the form shown as bipolar return to zero (BPRZ) gives more signal transitions, thus aiding bit (clock) synchronization. These forms of pulsed signals are known as line code formats.

5.113 Sequences of digital pulses correspond to binary digits, which in turn represent, for instance, alphanumeric characters that require another coding and decoding procedure. Certain applications require a number of bits to form what is termed a computer word or a data byte. A 7-bit character sequence is used to form the American Standard Code for Information Interchange (ASCII) code commonly used in computer-to-printer operations. This code is shown in Figure 5-6. Table 5-1 lists the abbreviations and their definitions.

Books have been written on various codes that contain many features, yet such features are not a concern to communications people as long as the type of line code is compatible with the transmission medium. Certain types of codes contain error detection and correction features, but the ASCII code is widely used in personal computer (PC) input/output operations and has undergone several revisions since it was first adopted in 1963.

Other alphanumeric codes are the old TTY (Teletype) code and the one developed by IBM called Extended Binary-Coded Decimal Interchange Code (EBCDIC), which is an 8-bit code with many characters. Error-checking is performed on the ASCII code using another bit, making a character 7 bits plus 1 parity bit for a total of 8. Bits 0 through 6 are the character's 7 bits and bit 7 is the parity bit. The EBCDIC code uses no parity bit.

Figure 5-4
Pulse digital signal

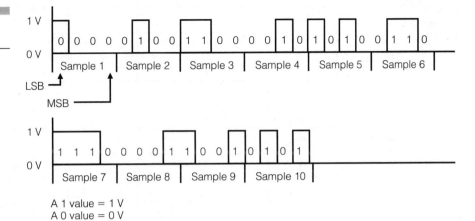

LSB

MSB

A 1 value = 1 V
A 0 value = 0 V

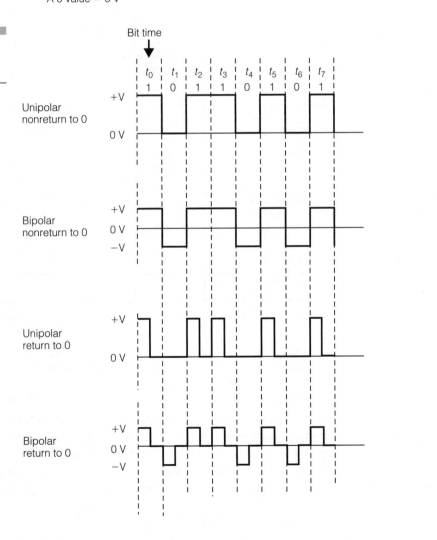

Figure 5-5

Some examples of
line-coding format

Figure 5-6
ASCII code bit/character chart

b4	b3	b2	b1		b7	0	0	0	0	1	1	1	1
					b6	0	0	1	1	0	0	1	1
					b5	0	1	0	1	0	1	0	1
	Bit number				Column								
b4	b3	b2	b1	ROW	0	1	2	3	4	5	6	7	
0	0	0	0	0	NUL	DLE	SP	0	@	P	`	p
0	0	0	1	1	SOH	DC1	!	1	A	Q	a	q
0	0	1	0	2	STX	DC2	"	2	B	R	b	r
0	0	1	1	3	ETX	DC3	#	3	C	S	c	s
0	1	0	0	4	EOT	DC4	$	4	D	T	d	t
0	1	0	1	5	ENQ	NAK	%	5	E	U	e	u
0	1	1	0	6	ACK	SYN	&	6	F	V	f	v
0	1	1	1	7	BEL	ETB	'	7	G	W	g	w
1	0	0	0	8	BS	CAN	(	8	H	X	h	x
1	0	0	1	9	HT	EM	)	9	I	Y	i	y
1	0	1	0	10	LF	SS	*	:	J	Z	j	z
1	0	1	1	11	VT	ESC	+	;	K	[	k	{
1	1	0	0	12	FF	FS	,	<	L	\	l	|
1	1	0	1	13	CR	GS	-	=	M	]	m	}
1	1	1	0	14	SO	RS	.	>	N	^	n	~
1	1	1	1	15	SI	US	/	?	O	_	o	DEL

SP, space; DEL, delete.

Several chip manufacturers make chip sets used for encoding and decoding both ASCII and EBCDIC codes.

5.114 Digital methods used for handling computer data and graphics are commonly used today in a variety of applications. The computer monitor has to display the ASCII-coded letters and numbers on a cathode-ray tube screen in its proper place and color. The monitor's digital circuitry has to contain chip circuitry that corresponds to a grid system of columns

Table 5-1

Definitions of ASCII
Abbreviations

| Column O | | Column 1 | |
Abbreviation	Definition	Abbreviation	Definition
NUL	All characters zero	DLE	Data link escape
SOH	Start of heading	DC1	Device control
STX	Start of text	DC2	Device control
ETX	End of text	DC3	Device control
EOT	End of transmission	DC4	Device control
ENQ	Enquiry	NAK	Negative acknowledge
ACK	Acknowledge	SYN	Synchronous idle
BEL	Attention alarm	ETB	End/transmission block
BS	Backspace	CAN	Cancel
HT	Horizontal tab	EM	End of medium
LF	Line feed	SS	Start/special sequence
VT	Vertical tabulation	ESC	Escape
FF	Form feed	FS	File separator
CR	Carriage return	GS	Group separator
SO	Shift out	RS	Record separator
SI	Shift in	US	Unit separator

and rows used to position the digital coded data. Because the ASCII data is used for an alphanumeric character, the computer and monitor program will place it on the next character position on the screen. This same program will give a screen background color along with the preprogrammed prompting messages. Again, the digital chip manufacturers have produced the necessary circuitry used to produce screen graphics.

This same screen technique is used to display digitally encoded television pictures. The specification given by some manufacturers for resolution is in dots per inch (DPI), which is an older printer-type specification for resolution. A more current specification is given in pixels, with 2048 × 2048 being the current specification for computer color monitors. Comparatively, this is far better than the resolution of most high-quality television sets.

Digital data values are often plotted on graphs or curves by digital plotters that require data handling by a computer. Under software control, the computer addresses the plotter with instructional information as to scale, size, and so on and feeds the plotter with digital data that is in turn plotted by a moving pen or a dot matrix. Manipulating data for graphic presentation is a big business in today's computer environment.

5.12 Digital Computing and Data Storage

PCs can be found in many homes in America, and the Internet, along with what is called e-commerce, is becoming commonplace. Computer manufacturers are producing PCs with faster speeds and huge amounts of memory. Large files can be stored on hard discs with humongous capacity. Programs can be purchased on a compact disc (CD), thus not requiring space on a hard drive. Most present-day computers found in homes and offices use many forms of storage such as CD-ROMs, diskettes, and solid-state random-access memory (RAM) chips. Modem units, either installed within the computer or as an external unit, allow the computer to be connected to the telephone (telco) system, which in turn connects to other online systems. It is here that the cable television system has more to offer in access speed than the telephone system. In brief, the coaxial cable drop has more high-speed data capacity than the copper-twisted telephone drop.

5.121   The heart of most PCs is the main microprocessor integrated circuit (IC) chip that controls the data transfer and computational speed. At present the fastest microprocessor operates at a clock speed of 2.8 GHz. Communications people, particularly those in the field of cable television, have a frequency upper limit of 750 or 850 MHz. This positions a cable system to communicate with computers at a high data-transfer rate. The main printed circuit board containing the main microprocessor chip is referred to as the motherboard. This circuit board also contains supporting circuits to the main microprocessor. Most of the electrical signals at 2.8 GHz occur in the microprocessor chip, that is, within the chip, thus limiting signal radiation to low levels. As microprocessor chip speeds increase, the input/output data rates increase as well. This requires more and more frequency bandwidth from the communications provider.

5.122   Transferring data from one workstation to another in a rapid manner is a requirement of many enterprises. In the business world, large data files are frequently loaded and unloaded between user computers and mass storage facilities. Many cable television systems operating today have the capacity to deliver high-speed data traffic to users connected to the system. Thus, the transfer of large quantities of data can take place a lot faster by cable operators with available channel capacity.

In the past, typical cable operators avoided commercial areas simply because they believed no subscribers were present and decided not to place their systems there. Cable operators interested in expanding their businesses by offering data-communication services to commercial enterprises will have to extend their plants to these areas. In short, they will go where the business opportunities exist.

At present, many banks, department store chains, and gas stations, to name a few, use very small aperture terminal (VSAT) systems to provide data-communication services. Most of these systems, however, operate in one direction with the upstream through the local exchange carrier (LEX). Cable systems should be able to provide such services to such customers.

5.13 Development of Data Communications

Providing high-speed data-communication services to commercial businesses should provide much added revenue to cable system operators willing to upgrade and expand their distribution systems. Many cable systems already have been exposed to digital signal transmission from the use of addressable converters. Data communications using digital signal transmission will be used with many of the new cable modems for HDTV and pay-per-view program selections. The progressive cable operator should install optical fiber in both the downstream and upstream path along with standby power supplies needed to increase system reliability.

5.131 Soon after PCs appeared as office workstations, it became apparent that computer systems would need to be interconnected. The first solution was to use a dual-tone multifrequency (DTMF) method. This method resulted in the acoustic coupler modem, in which the telephone handset was placed in a cradle that would translate digitally encoded data into DTMF tones. Transmission could take place in both directions from the mouthpiece and the earpiece. Of course, data communications were slow.

Next, the separate modem system was developed using frequency shift keying (FSK). The modem converted the digital data pulses into two frequencies representing ones and zeros. FSK data received by the telco network was converted to pulsed signals and sent to the receiving computer via its input/output port. Figure 5-7 illustrates modem development.

At present, most modems are plug-in cards inserted into the computer main circuit board (motherboard) and connected directly to the telephone system via a modular telephone connector plug. As modems were required to transfer data at higher bit rates, the complexity of the modulation schemes increased. From simple FSK modulation with a data rate of 300 to 600 bits per second, development led to binary phase shift keying, which increased the effective bit rate to 1200 bits per second (bps). Quadrature amplitude modulation (QAM) increased the telco modem effective bit rate to the present standards. Current modem advertisements state the modem bit rate at 56 Kbps, qualified by the statement that the speed at 56 Kbps depends on the line quality of the local exchange carrier (LEX). Because a cable television system could easily provide data rates equal or greater than this speed, the development of a cable modem is necessary for a cable operator to enter this phase of service.

5.132 The use of telco modems allowed residential and business computer systems to interconnect. A network of simple dial-up modems and leased line systems produced an area network among connected users. In-house networks where computers were connected via buses, switchers, and channel routers formed local area networks (LAN).

Figure 5-7
Development of the
telephone-computer
modem

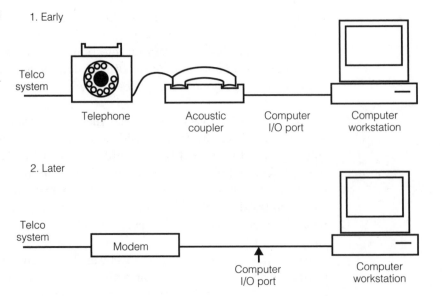

5.2 Present Data Communications Systems

Present-day data networks consist of networks of owned intracompany networks connected to national and worldwide telephone systems that form a commercial telecommunication network. LANs are connected to wide-area networks (WANs) to various interconnected providers, including the telephone systems and other common carrier operators. It is here that progressive cable operators can provide a necessary and useful service to the telecommunications industry.

As noted in Chapter 1, the principal professional organization representing the cable television industry changed its name to the Society of Cable Telecommunication Engineers. This was indeed an appropriate action. Cable systems should be and are aligning their operations to provide not only television programming but also computer interconnection services. Cable television operators, in the not-too-distant future, will be an Internet provider (IP) made possible by a cable modem. Because the Internet offers voice service, cable operators acting as IPs also provide a form of telephony.

5.21 LANs and Topology Types

LANs were developed as a necessity for supplying data connections required by the commercial business world. Because no standards were specified, technicians and engineers made rudimentary networks to solve particular problems. Network topologies were finally developed and several types of LANs evolved.

These network topologies exist in a variety of forms, such as a star network, where all end-user stations pass through a main switching center. Other topologies are the tree-branch network, the ring-type network, and the bus network. These topologies are shown in Figure 5-8.

Various techniques are also employed to control the network. A central switcher controls the star-type network. For the tree network, each station has its address and control from branch to branch through the main station. A bus network is controlled by a collision-detection system used to prevent all stations from trying to communicate at once. The ring topology usually uses a token-passing scheme in which the station that needs to transmit must possess the token.

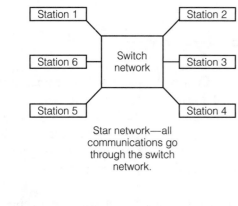

Star network—all communications go through the switch network.

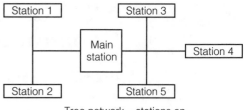

Tree network—stations on same branch communicate; those on different branches go through the main station.

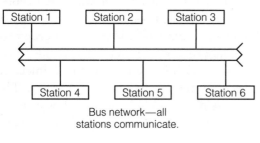

Bus network—all stations communicate.

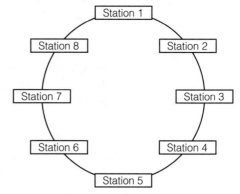

Ring-network—stations act as repeaters for each other.

Figure 5-8
Network topologies

Each network topology has its advantages and disadvantages. The specific application often determines the one that will be best suited. A brief study of some of the more popular LAN systems follows in the next sections.

5.211 Ethernet is one of the oldest LAN types still in existence today and it seems as if it keeps reinventing itself. Basically, Ethernet is a bus topology and is one of a family of IEEE 802 architectures. Ethernet uses coaxial cable that supports data rates of 10 Mbps. This cable has a 50-ohm characteristic impedance. Ethernet rates over coaxial cable have been pushed to 100 Mbps, while for shorter distances over 10 Mbps copper twisted-pair can be used.

Recently, as the need for faster transfers of large data files has become critical, Ethernet systems converted to optical fiber. Thus, for distances of 200 m, multimode fiber systems are adequate, while single-mode operation distances of approximately 5 km are attainable at speeds up to 1 Gbps. The conversion to fiber was quite simple, following the same routes and often not requiring any electronics between stations.

Ethernet was originally developed through a concerted effort by Xerox, Intel, and DEC. Workstations operating on the bus cable had either an Ethernet modem or a network interface card (NIC) installed into the workstation computer. Ethernet operates on the bus using carrier sense multiple access with collision detection (CSMA/CD). A station wishing to transmit has in the modem or NIC a circuit that looks for a data carrier on the bus. Seeing none, it transmits and if at that same moment another station transmits, the collision circuit in each of the transmitting station's NIC will call for a time-out. The length of the time-out period varies between users. Once a connection is made between stations and a carrier is present, it prevents other stations from trying to transmit.

For many businesses, Ethernet LANs provide the needed data transfers between computer workstations. Many Ethernet cable systems used a piercing pressure tap similar to early cable television systems that was used to connect the data transceivers to the bus cable line. An example of an Ethernet system is shown in Figure 5-9.

The Ethernet LAN is essentially a transmission line system and should be properly terminated. Branching has to be made with coupling networks

Figure 5-9
Ethernet bus system

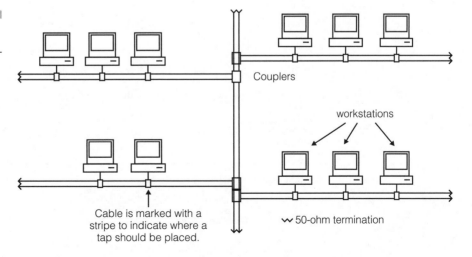

all properly impedance-matched. Repeater amplifiers can be placed in the line when needed to increase the carrier signal level. The data rate, or bit rate, is related to the data packet size by the transmission path length. For short distances of approximately 100 meters, 10 Mbps can be supported by unshielded twisted dual-pair cable, which is actually telephone cable. This is very attractive for small intrabuilding LANs using low cost, easily installed telephone wire. In many cases the business occupants found excess telco type wire already installed and in place. Thus, a low-cost 10 Base-T LAN could be quickly put together. Table 5-2 lists the generic progression of the Ethernet type LANs.

Ethernet software requires that data be sent in packets with the appropriate header containing the addresses of the sender and receiver. The formation of a data packet is shown in Figure 5-10. The preamble section is often 8 bytes of zeros and ones, providing enough transitions to establish bit synchronization (clock sync). Following the preamble is the destination address of 6 bytes. This address can be a specific station address or a broadcast for every station to receive or a selected group of stations to

Table 5-2

Some Examples of
Ethernet Local Area
Networks

Ethernet Type	Data Rate (in Mbps)	Segment Length (in m)	Cable Type	General Designation
10 Base-5	10	500	Coax RG-8 50-ohm	Thicknet
10 Base-2	10	200	Coax RG-58 50-ohm	Cheapnet
10 Base-T	10	185	#24 4 wiretelco	
100 Fast Ethernet	100	100	50-ohm coax	
Gigabit Ethernet	1000	25	50-ohm coax	
1000 Base-Lx	1000	220	MM fiber[a]	
1000 Base-Lx	1000	550	MM fiber[b]	
1000 Base-Cx	1000	25	Copper twisted pair	
1000 Base-Sx	1000	5000	SM fiber[c]	

[a]Multimode 62.5/125 mm fiber optical cable

[b]Multimode 50/125 mm fiber optical cable

[c]Single-mode 5/125 mm fiber optical cable

Preamble 8 bytes	Destination address 6 bytes	Sender address 6 bytes	Field type 2 bytes	Data field 46–1500 bytes	CRC 4 bytes

Figure 5-10
Ethernet packet

receive the enclosed message. The field type consists of 2 bytes describing the data field of 46 to 1500 bytes. The remaining 4 bytes are the cyclical redundancy-checking sequence.

The line code for Ethernet is what is called the Manchester code at the usual 10 Mbps. This code assures a pulse transition with every bit, thus maintaining clock synchronicity.

5.212 Another type of LAN is the token-ring network, as shown in Figure 5-8. Each workstation is placed within the ring, where the cable passes through each station. When no station is transmitting, the ring rotates around from station to station. The token is actually a coded bit sequence. When a station is ready to transmit to another station, it captures the ring and then proceeds to transmit its data. Data is in the form of a packet of data bits. The transmitting station has to send data packets of specific sizes to fit the allowed transmitting time slot, as determined by the LAN specification. The receiving station receives the token and returns it to the sender upon completion of the data packet transfer. The sending station then gives up the token to be circulated again, making it available to another sender.

Variations of the token-ring LAN depend on what equipment the manufacturer provides. Naturally, most providers offer the usual bells and whistles, such as station priority levels and packet sizes.

Ethernet and token-ring LANs have quite a few pros and cons. Substantial delays can take place, particularly for large rings because the token takes time to circulate. The ring delay time can cause a loss of time equaling a large number of data bits. As the ring becomes loaded with data, a station can wait a long time before it seizes the token.

Token-ring LANs appeared around 1980 and gained acceptance rapidly because this type of network had fewer data-congestion problems than Ethernet LANs. The IBM Corporation introduced its version of token-ring architecture around 1985 (IEEE Std 802.5). Today many token-ring LANs are operating, providing adequate service as required by users.

The cable used in token-ring networks is easily recognized as shielded twisted-copper pairs with a two-prong screw-on connector. Data rates of 4.0 Mbps to 10 Mbps are attainable on token-ring LANs.

5.213   Using cable television technology, a broadband LAN was developed. This type was referred to as manufacturers automated protocol/ technical office protocol (MAP/TOP). This type was developed through the effort of General Motors and various equipment providers. MAP/TOP was needed badly on the manufacturing floor for automobiles. The production line is a huge electrical-noise generator, making twisted-pair and even coaxial cables or shielded-pair cable too full of noise to properly carry data. Because automatic or robotic manufacturing techniques require data communications with the controlling computer, some quiet communication methods were developed. MAP/TOP modulates high-frequency data carriers with digital data where electrical noise is less, in the high frequency spectrum.

By stacking multiple carriers, as in cable television systems, large amounts of data could be handled by MAP/TOP. Using 75-ohm coaxial solid-aluminum cable, the usual passive circuits (such as taps, splitters, and couplers) and line-repeating amplifiers provided the needs of the design. Since data had to flow in both directions, a frequency band was assigned to the upstream direction and the downstream direction. An example is shown in Figure 5-11.

In this figure, the upstream band is 5 to 174 MHz for a bandwidth of 169 MHz, and the downstream band is 234 to 400 MHz for a bandwidth of 166 MHz. This allows for nearly equal upstream and downstream bandwidth. As more data channels are needed for the downstream direction, the upper frequency limit can be increased to accommodate the need.

The MAP/TOP LAN using coaxial cable television technology required a lot of maintenance, including testing for signal leakage in the air safety frequency bands. Also, such a network consumed quite a lot of power and made keeping some of the equipment cool a priority. Many of these systems have been converted to fiber-optic networks.

5.22 Fiber-Optic Systems

Fiber-optic cable, as discussed earlier, has less optical signal attenuation than coaxial cable has for electrical television frequency attenuation. Thus, for most common physical cable distances, as found in commercial

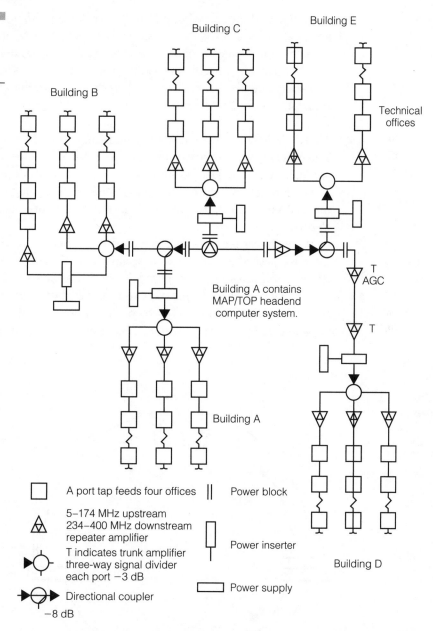

Figure 5-11
Example of a
broadband MAP/
TOP-type LAN

Building C

Building E

Building B

Technical
offices

Building A contains
MAP/TOP headend
computer system.

T
AGC

T

Building A

Building D

☐ A port tap feeds four offices   ‖ Power block

△ 5–174 MHz upstream
234–400 MHz downstream
repeater amplifier

▶○ T indicates trunk amplifier
three-way signal divider
each port −3 dB

▶○▶ Directional coupler
   −8 dB

▯ Power inserter

▭ Power supply

Downstream signals containing modulated carriers with digitally encoded manufacturing instructions make up the forward signal. Messages are sent from the headend to the manufacturing machines and offices.

Upstream signals on modulated RF carriers contain data, requests for materials, and schedule progress. These are sent through the headend to other offices and warehouses.

buildings and industrial parks, no electronic amplifier or signal repeaters are needed for a fiber-optic system. Therefore, most optical fiber systems are completely passive except for the terminating transmitter and receivers. As business computers sped up, the requirements for high-speed data transmission also increased, and fiber-optic development provided the solution.

5.221   One such system was the fiber-distributed data interface (FDDI). The American National Standards Institute (ANSI) developed the standard X3T9.5 for the FDDI system that operates at 100 Mbps using a counter-rotating ring topology. As in most ring network topologies, a token-passing method is used to gain access. The FDDI network topology is shown in Figure 5-12.

It was necessary to develop a special connector for the FDDI network. Because both the primary and secondary fiber ring passed through each node, a dual fiber connector was developed as shown in Figure 5-13.

An FDDI node operates as a repeater of data that is not addressed to that particular station. This technique allows for distances up to 2 km between node stations. Many FDDI networks use a multimode-graded

Figure 5-12
Dual-ring FDDI network

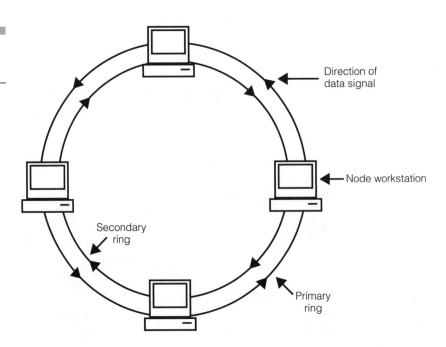

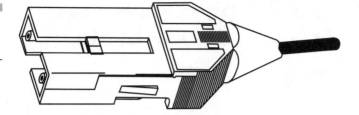

Figure 5-13
FDDI dual-fiber
connector

index fiber of the 62.5/125 mm variety, but this is not a hard and fast rule. Other fiber types can also be used. For most FDDI applications, LED light source transmitters and pin-photodiode receivers are adequate. For larger ring sizes with nodes farther apart, single-mode fiber and laser transmitters are used. Often large corporate network LANs will use FDDI dual ring networks as a backbone connecting other regional LANs. The interconnecting device is referred to as a bridge, which communicates between different network types.

5.222 Another network that uses optical-fiber technology is the synchronous optical network (SONET), which was developed recently for the telecommunications industry. The SONET standard is for North American synchronous rates, and a similar standard for Europe is the synchronous digital hierarchy (SDH). SONET/SDH systems operate in the digital domain, are synchronous, and operate on single-mode fiber optical networks. The digitally encoded data is packet-switched and is organized into 810-byte frames, which include headers containing signal routing information. Data frames can be switched in and out individually without unbundling the SONET frame. The data transmitted using SONET technology can be digitally encoded telephone traffic and commercial data traffic, all intermixed on SONET frames. The standard specifies frame sizes and data rates, as well as configures a ring topology that provides a redundant communication path. Thus a SONET system is essentially self-healing, providing a very reliable network. Data rates for SONET systems are all multiples of each other, thus preserving the synchronization. SONET systems will be revisited in other sections of this book. Cable operators and their technical personnel should look at the possibilities of providing access to all these data carriers as new business opportunities and the accompanying source of revenue.

5.3 Telephone Systems and Digital Technology

The largest and oldest telecommunication system in the United States is the telephone system. Since its invention by Alexander Graham Bell, the telephone network has continuously grown and progressed technologically. The telephone system is known for its high quality and reliability, setting high standards for other industries to follow. The telephone industry, mainly American Telephone & Telegraph (AT&T), produced many useful inventions and scientific discoveries through research by Bell Labs on speech, hearing, circuit switching, and electronic components. Such efforts produced valuable knowledge for the benefit of industry and the public in general.

5.31 Basic Telephone System

The basic telephone system consisted of copper pair cables installed on utility poles strung about the countryside, connecting the telephone sets in subscribers' homes to the central terminal location called the local exchange (LEX). This network remained essentially an electromechanical system for many years. When electronic amplifiers became practical and radio methodology was developed, the telephone companies applied these technologies to long-distance telephone service.

Presently, the telephone network is rich in fiber optics, which has essentially taken over the long-distance service and provided other services to commercial enterprises needing data communications. The cable television operator stayed out of commercial areas where the telephone systems were providing telephone communications. Oddly, the telephone companies expanded their fiber-optic plants into commercial areas and not into residential areas. The cable operator remains today with the residential subscriber coaxial cable drop with nearly a gigahertz of bandwidth, while the telephone company has its twisted-pair drop wire of limited bandwidth.

5.311 The basic telephone set or instrument has grown through many stages of development. The rudimentary telephone network consisted of copper twisted-wire pairs insulated with silk, wax, and wax paper covered with a lead jacket. Local service was only provided initially. Each instrument or telephone set had a mouthpiece (transmitter) and an earpiece (receiver) with a switch hook to hold the earpiece. All calls initiated by lifting the switch hook connected to the manually operated local exchange office, where a live operator requested the called party's number. A plug on the switchboard was placed into the jack corresponding to the called party's number, and the operator would turn on the ringing current.

Eventually, the operator and manual switchboard was replaced by a dial system that automatically switched the calling station to the called station's circuit and performed the ring, connect, and disconnect functions. The rotary dial was a serial pulse generator that operated the electromechanical (relays) circuit switch at the LEX. Long-distance or toll service to other LEXs was accomplished by trunk lines operated by toll switches. At first, these toll switches were manually operated by a switchboard operator who made the circuit connect and disconnect, keeping track of time and charges. Later the toll service was operated by a switching system similar to that used at the LEX.

These early telephone systems were troublesome, mainly the outside cable plant. Breaks and cracks in the lead jacket allowed moisture to enter and soak the fabric insulation, causing cross-talk and false ringing. Specially trained cable splicers who could work with lead and hot solder were needed to service and extend the cable plant. The arrival of plastic insulated and jacketed cable essentially solved the lead cable problems.

Early telephone sets were electromechanical in design and construction. The hybrid, as it was known, was a series of transformer phases wired to prevent transmitted voice currents from blasting the earpiece and incoming received voice currents from being dissipated in the transmitter. Telephone personnel referred to the hybrid as a two- to four-wire converter. The hybrid concept is shown in Figure 5-14. The block diagram of the generic dial telephone is shown in Figure 5-15.

With the development of touch-tone service, the push buttons selected combinations of two tones for each button. These tones were at first generated mechanically (form of chimes) and progressed to a series of electronic oscillators. The electronic telephone is what we have today. A block diagram of such a telephone is shown in Figure 5-16. The action of the hybrid circuit is synthesized by an integrated circuit (IC) chip.

5.312 The telephones in a local community all go to the LEX office by way of twisted-pair plastic insulated cables installed as buried plant or overhead utility poles. These wire pairs are connected to a cross-bar switch that can be an electromechanical relay switch bank or a solid-state microprocessor-controlled switch bank.

Figure 5-14

Transformer-type telephone hybrid circuit

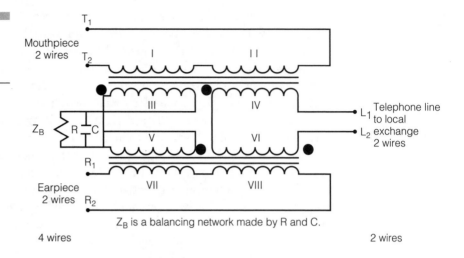

Z_B is a balancing network made by R and C.

Figure 5-15

Basic telephone set

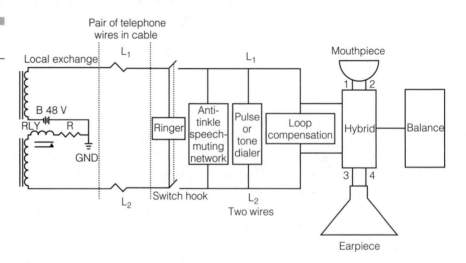

The LEX is connected to other LEX offices through an interoffice trunk line. Long-distance toll calls are connected to toll center offices that connect to intertoll trunks. Formerly, the interoffice and intertoll cable systems were copper, but most have been replaced by optical fibers. Regardless of the method, the telephone system network topology is the same. This interconnection topology is shown in Figure 5-17.

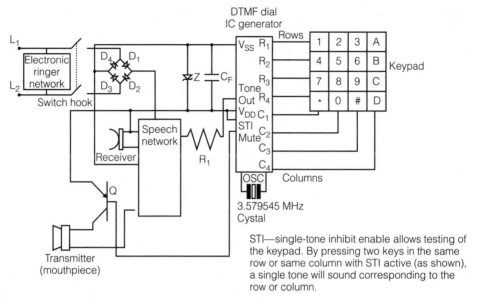

STI—single-tone inhibit enable allows testing of the keypad. By pressing two keys in the same row or same column with STI active (as shown), a single tone will sound corresponding to the row or column.

Diode bridge consisting of D_1, D_2, D_3, and D_4 protects against wrong polarity line connections and feeds DC power to the speech network and DTMF generator. Z is a zener-diode surge protector. C is a filter capacitor. Q acts as a muting control for the transmitter.

Figure 5-16
Electronic telephone set

There are five principal classes of telephone system offices with the LEX office or end- user office. Long-distance telephone service can involve all five of the office classes interconnected by preferred and alternate routing. The hierarchy of a call between two parties using all classes of offices is shown in Figure 5-18. The classes of offices are supervisory in nature and have control over certain areas. The Class 1 or regional office is the highest class, and there are only 10 or 12 in the United States and two in Canada. There are more of the lower class offices with the Class 4 office at the local level. The Class 5 office has the greatest number of offices.

5.313 The telephone network in this country has contributed to the success of many companies using the services. The network has grown from many small regional exchange offices into a huge network providing wired

Figure 5-17
Telephone-switching
topology

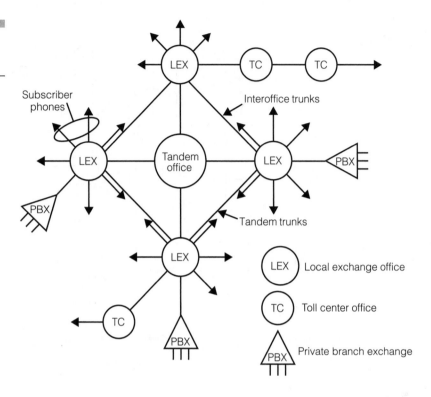

communications for voice, data, and some video services. Actually, the system as it is today is a network within a network.

Today telephone companies use a supervisory network named signaling system 7, or SS7, to manage the service network. This system is a continuation of the common channel interoffice signaling (CCIS) network. The SS7 uses a packet-switching protocol to send call processing and status information that is essential to proper and accurate billing system charges to the subscribers and end users. The switching functions are performed by electronically controlled switches, referred to as 4ESS or 5ESS switches. The 5ESS switch is used when service demand is low, and 4ESS switches operate in a fully configured mesh-type network, providing many alternate paths needed to process a high volume of calls. Instructions setting up the call routes are sent by the SS7 to the 4ESS switches as the numbers are being dialed by the calling telephones.

Still, the largest providers of telephone service are the Bell system companies, often described as the regional Bell operating companies (RBOCs). Other telephone providers operate in a similar fashion and all have access

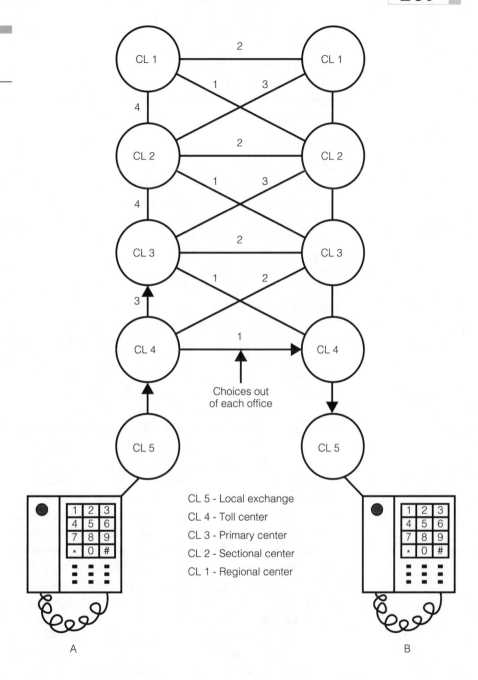

Figure 5-18
Telephone-system
call-office hierarchy

CL 5 - Local exchange
CL 4 - Toll center
CL 3 - Primary center
CL 2 - Sectional center
CL 1 - Regional center

to one another's systems, thus providing proper seamless services to the businesses and general public.

The telephone network as we know it today grew from a simple copper wire connected system to one consisting of coaxial cables, microwave radio, satellite, and optical fibers. These technologies are used for long-distance and trunk routing, while the copper cables deliver telephone service the last mile.

5.314 The trunk long-haul telephone services are processed over mostly optical fiber installed over the last 10 to 15 years. Switching functions performed by the 4ESS electronic switch are not simply connecting wires and cables together but connect the communication path. Because the telephone system converted to digital technology, telephone traffic consists of digital data words or bytes. Many calls are bundled together into digital transportation streams transmitted along trunk routes to the call destination, where the bytes are stripped out, converted to analog speech, and sent to the called party's telephone.

It is interesting to note that early telephone long-haul trunking was done by coaxial cable. Analog voice signals were converted to higher frequency bands by single sideband radio techniques. One telephone one-way speech signal would occupy the upper sideband of a suppressed carrier and another, the lower sideband. At the receiving end, the carrier was inserted and each sideband was demodulated.

Coaxial cables operating for long distances provided many telephone channels and cascaded amplifiers acting as signal repeaters were used along the cable run. Cascaded amplifier theory was put into practice and was the forerunner to cable television systems. The multiplexed signal scheme is shown in Figure 5-19. When no rights-of-way were available for cables, these signals could be converted to microwave radio frequencies and were transmitted the required distances. Converting back and forth between the cable signals and microwave radio techniques contributed to the transcontinental long-haul telephone system. It's possible that even today some of these type facilities are still in use.

For quite some time now, the telephone companies have been installing large quantities of fiber-optic cable that is being used in long-haul trunk line applications. Also, the SS7 is nearly all optical fiber. Digital data communication traffic is transported mainly by SONET technology. The drop/add service at the various exchange offices is handled quite well by SONET. Actual switching is performed by 4ESS by merely switching data bytes in and out of the SONET data payload. Further discussion of SONET technology appears in Chapter 7.

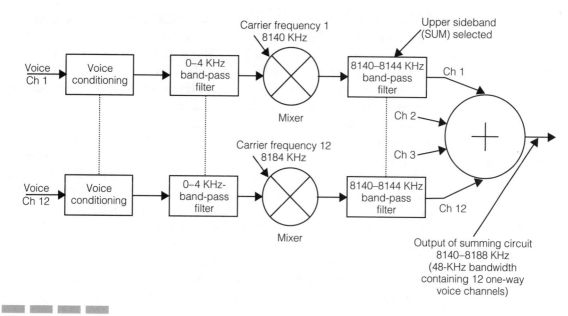

Figure 5-19
Coaxial cable with 12 voice channels

5.32 Digital Telephone Methods

Shortly after World War II, the telephone industry converted to digital technology based on cost and system efficiency factors. This was a timely and appropriate decision to make. Because switching theory is actually a binary or digital subject, it is a normal extension of the technology to apply to the voice signal as well. Thus, the T-1 system was introduced. Because some telephone engineers use the word *carrier* to describe the message transportation method, the T-1 digital system is called the T-1 carrier, which is presently classified as digital system-1 (DS-1). Expansion of DS-1 speeds by time-division multiplexing (TDM) methods increased bit rates to DS-2 and DS-3 levels.

5.321   The T-1 system digitizes 24 analog telephone channels into a data stream of bits. Because each telephone analog channel of 4 KHz is to be sampled, according to Nyquist's sampling theorem, each of the channels has to be sampled at 8000 samples per second. Once each of the 24 channels has been sampled, a framing bit is needed to indicate the end of the frame.

Each sample of the analog data needs 8 bits to accurately describe the signal amplitude. The net result is that 24 channels (one way) of the telephone service give a bit rate of 1.544 Mbps. This is shown in Figure 5-20. This method is known as pulse code modulation (PCM).

Also, because each telephone channel has a time slot in the bit stream, it is also referred to as TDM. The T-1 system can be carried on a four-wire system for full-duplex transmission. It might seem impossible that copper twisted pair can carry a pulsed signal operating at 1.544 Mbps. This line rate is actually transmitted long distances as a carrier phase amplitude−modulated signal where several bits are contained by one phase and amplitude value. As long as the physical transmission wire or cable system can carry the carrier and preserve the modulation intact, the effective bit rate can be realized. Wider-band cables, such as coaxial cable and fiber-optic cable, with higher bit rates can be transmitted. These higher bit rates are achieved by multiplexing the basic T-1 carrier.

Channel bank equipment either at the LEX or a private branch exchange (PBX) at various commercial locations can output higher order

Figure 5-20

Formation of T-1 systems (DS-1)

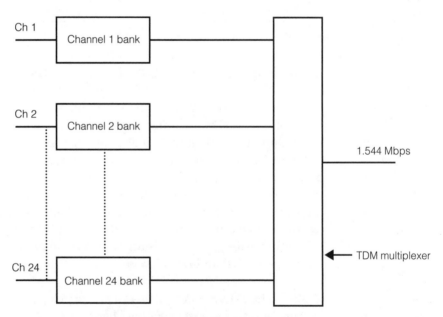

Each channel is sampled at 8000 frames/second providing 8 bits/sample which is 64,000 bits/channel/second.
For 24 channels, plus 1 bit (framing bit):

64,000 bits/channel/second × 24 channels/frame = 1,536,000 bits/second, plus 8000 frame bits = 1,544,000 bits/second = 1.544 Mbps

Example 5-1

For basic T-1 system:
DS-1

$$\frac{8 \text{ bits}}{\text{channel}} \times \frac{24 \text{ channels}}{\text{frame}} = \frac{192 \text{ bits}}{\text{frame}} + \frac{1 \text{ frame bit}}{\text{frame}} = \frac{193 \text{ bits}}{\text{frame}}$$

At a sampling rate of 8000 frames/second:

$$\frac{193 \text{ bits}}{\text{frame}} \times \frac{8000 \text{ frame}}{\text{second}} = 1.544 \text{ Mbps}$$

For basic T-1 C:
DS-1C

$$\frac{8 \text{ bits}}{\text{channel}} \times \frac{48 \text{ channels}}{\text{frame}} = \frac{384 \text{ bits}}{\text{frame}} + \frac{10 \text{ sync bits}}{\text{frame}} = \frac{394 \text{ bits}}{\text{frame}}$$

At a sampling rate of 8000 frames/second:

$$\frac{394 \text{ bits}}{\text{frame}} \times \frac{8000 \text{ frame}}{\text{second}} = 3.152 \text{ Mbps}$$

T carriers. Present-day channel banks using large-scale integrated circuit (LSI) technology work in full duplex where the analog-transmitted speech signal is digitized, compressed, and transmitted. The return digital signal is also decoded, decompressed, and converted to analog, providing full-duplex service. Channel banks that operate at higher speeds can stack basic T-1 carriers into higher order line speeds. For example, two T-1 frames consisting of a total of 48 channels will produce a line speed of 3.152 Mbps. This method is illustrated in Example 5-1. For short transmission distances of under 1 mi, data can be sent over copper wires using a +3 V and−3 V bipolar format called alternate mark inversion (AMI). This format provides improved signal synchronization due to increase of transitions in the signal. The digital signal hierarchy developed from the basic T-1 system is shown in Table 5-3.

5.322 The telephone industry in both the United States and in Europe has expanded rapidly into digital technology. Basically, the T-1 digital carrier consisted of 24 analog telephone channels. Each channel is sampled at 8000 samples per second into 8 bits per sample, giving the basic DSO data rate of 64 Kbps. This DSO rate can support computer digital data. Therefore, a 24-channel T-1 system can be subdivided into, for instance, 18 voice channels (18 DSOs) plus 5, 56-Kbps channels (5 DSOs) and 3, 9.6-Kbps data channels can be multiplexed to 1 DSO channel, therefore completing the 24 DSO channels. Using this method, commercial digital data services can be mixed into T-1 frames and transmitted to offices in various cities. Banks and financial institutions that require high-speed data communications use the telephone network facilities to carry their business data traffic intermingled with digital telephone traffic.

Table 5-3

The Digital Signal
Hierarchy

Level	Number of Voice Channels	Bit Rate (in Mbps)	Transmission Medium
DS0	1	64[a]	Copper twisted pair
DS1	24	1.544	Copper twisted pair cable, coaxial, or radio
DS1c	48	3.152	Radio, coaxial cable, optical fiber
DS2	96	6.132	Radio, coaxial cable, optical fiber
DS3	672	44.736	Radio, coaxial cable, optical fiber
DS4	4032	274.176	Radio, coaxial cable, optical fiber

[a]In Kbps

Packet-switched networks, which could carry both voice and data transmission, were developed in the United States in the 1960s. The original network consisted of nodes connected together over a wide area of leased lines. The nodes acted as switching centers. The terminal devices, known as data terminal equipment (DTE), transmitted messages to other DTEs. The early method operated as a datagram system sending small information packets called datagrams. As time went by, an agreement between the United States and Europe through efforts of the International Standards Organization (ISO) adopted the X.25 standards for packet switched network interfaces. This concept used the first three layers of the open systems interconnect (OSI). The OSI concept consists of a seven-layer protocol and is illustrated in Figure 5-21.

The main transport system for X.25 packet networks is done through leased telco lines with a direct dial-up backup. The packet networks developed for business data traffic work well in North America and Mexico. The DTEs used to be X.25 compatible and use the standard 15-pin connector. The X.25 packet also has to carry the appropriate flag, address, and control fields, followed by header and data fields. At completion of the three protocol layers of the OSI model, a flag follows the frame-check sequence character. Other network types requiring access to an X.25 packet network require the use of a piece of equipment called a bridge. The transmission speed depends on the specifications of the lines making up the physical layer, and it is this type of communication service that can be developed by enterprising cable television systems.

Figure 5-21
Open-system
interconnect
7-layer model

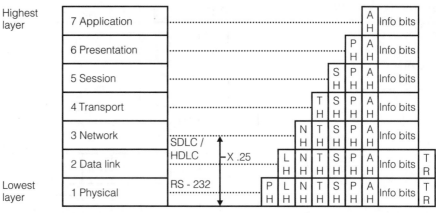

All headers needed if all
seven layers are addressed.

5.33 Fiber-Optic Methods

As stated earlier, the telephone industry in the United States and Canada
has installed thousands of miles of optical fiber, essentially replacing trunk
lines and long-haul lines. Voice and data can be transmitted at extremely
high speeds over long distances. Telephone traffic and data arrive at different nodes, all with various rates and without being synchronized. Thus,
the asynchronous transfer mode (ATM) was developed to carry voice and
data in the same packet. ATM accommodates the dispersion of voice packets and data packets better than a TDM or an ordinary packet-switched
network. For TDM, each channel has a time slot assigned that makes up
a frame. To intermingle voice and data into 8-bit octets, a voice channel
will be assigned alternate time slots with data. When no data is present
for a time slot, an idle octet is inserted. Packet-switched networks usually
alternate between voice and data traffic. Both TDM and packet-switching
can cause objectionable latency or delays for voice channels.

5.331 ATM is more efficient because no idle octets are used and all time
slots can be used to carry data or voice traffic. This concept is shown in
Figure 5-22. ATM makes more efficient use of bandwidth than other techniques and has the least delay times. The ATM cell contains 53 octets:
5 octets make up the header, and 48 octets compose the payload. An octet
is an 8-bit group, often referred to as 8-bit bytes, where 2 bytes would
make 16 bits. Most system data conforms to multiples of octets that start

Figure 5-22
Comparison of TDM, packet switching, and ATM data formats

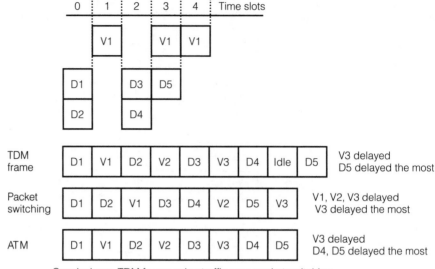

Figure 5-22
Comparison of TDM, packet switching, and ATM data formats

Conclusions: TDM favors voice traffic over packet switching.
TDM wastes time slots on idle.
ATM makes efficient use of bandwidth.

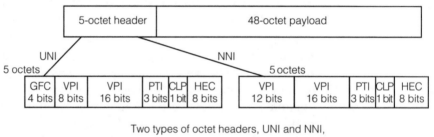

Figure 5-23
Format for an ATM cell

Two types of octet headers, UNI and NNI,

UNI Header	NNI Header
GFC generic flow control	VPI virtual path identifier
VPI virtual path identifier	VC2 virtual channel identifier
VC2 virtual channel identifier	UNI user network interface
PTI payload type identifier	PTI payload type identifier
CLP cell loss priority	CLP cell loss priority
HEC header error control	HEC header error control
UNI user network interface	NNI node network interface

with the 8-bit data word or byte and goes on up to 64 bits. User data is parceled into a 48-octet payload portion of the cell while the header portion contains control, routing, and switching information. The ATM cell format is shown in Figure 5-23.

ATM is a form of packet-switched network technology where a cell is a highly defined fixed length unit of data. User data is not fitted into a particular time slot. The user just keeps placing ATM cells on the transmission facility until the message is complete. The cell headers contain error control functions, making sure the header information is valid. The transmission facilities have installed ATM-type switching and routing equipment that act on the cell header information to control the flow of the cells.

Telephone companies carry ATM cells for customers using mostly fiber-optic cables. For longer distances, ATM cells are easily carried on the high-speed SONET because the cell structure is completely compatible with SONET requirements. SONET, as developed in North America, is compatible with the synchronous digital hierarchy (SDH), a European standard. As shown in Figure 5-24, the common point of connection between SONET and SDH is the optical carrier 3 (OC3) and SDH synchronous transfer mode (STM) level, which is three times the OC1 line rate. The SONET/SDH systems can handle the telco TDM digital signal (DS) formats with the DS-3 the entry level. The DS-3 line rate of 44.736 Mbps (T-3 carrier system) can be fitted to the OC1 slot with the necessary overhead.

There are many methods and specifications for transmitting data through communication networks. ATM is especially useful in gathering asynchronous data from various tributaries and forming data cells that can be applied to SONET data payloads. In essence, nonsynchronized

Figure 5-24
SONET/SDH line-rate
formation

SONET	SDH	Line rate (Mbps)	Multiplier
OC1		51.84	
OC3	STM-1	155.52	3 × OC-1
OC9		466.56	9 × OC-1
OC12	STM-4	622.08	12 × OC-1
OC18		933.12	18 × OC-1
OC24	STM-8	1244.16	24 × OC-1
OC36	STM-12	1866.24	36 × OC-1
OC48	STM-16	2488.32	48 × OC-1
OC192		9953.28	192 × OC-1

data is transported in cells that are transmitted synchronously. Also, digital voice traffic latency or signal delay is not tolerated well. Customers do not want their voice conversations containing delays. ATM, as stated earlier, is the least objectionable as compared to a TDM or packet-switched method. Data, of course, can tolerate latency because receivers will wait until the whole message is received intact before transferring to the user for processing.

5.332 The SONET system specifications and cell structure were developed by Bellcore and are considered as the high-speed toll road of the information highway. As developments occur in optical-fiber transmission, data transfer rates have increased to the equivalent of 320 Gbps versus dense wavelength division multiplexing (DWDM) technology. The SONET transmission system has been shown to be like passengers on a commuter train. The first building block for SONET transmission is the STS-1 level, which consists of header information plus payload data. The STS-1 basic building block is shown in Figure 5-25.

The train similarity indicates the synchronous manner, or the connected-together methodology of the SONET system. Synchronism must

Figure 5-25

Synchronous transport signal (STS-1) basic building block format

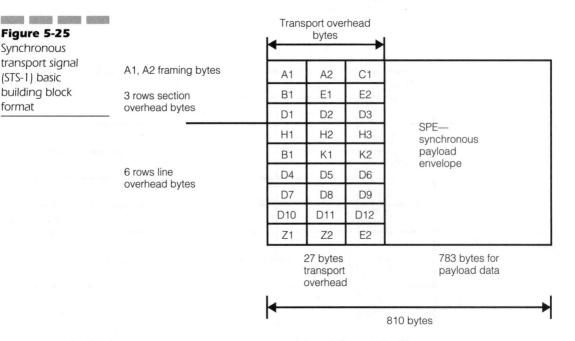

be maintained through the signal transport medium so the data slots can be unloaded and loaded with drop and add data riding the train. Scrambling the data bytes in the payload information provide enough signal transitions to aid in clock recovery. SONET-based systems can provide large amounts of high-speed digital transmission carrying voice, data, and video traffic.

5.4 Digital Video and HDTV

The digital transmission of television in some parts of the country is operational, and test data is being recorded and analyzed. Broadcasting in the digital domain has been studied and written about for a long time. The FCC finally accepted the transmission standard of 8-VSB along with all the base-band digital signal specifications necessary to implement service. Now, broadcasting trials are under way.

Some questions remain, such as how does a fading signal or multipath interference affect transmission, and will many people need an outside antenna to obtain proper signal level? From the consumer point of view, how much will a decent digital television receiver cost and will it have a flat, wall-mounted screen or be contained in a huge box the size of a piano? Also, will the broadcasting stations transmit in a one-channel HDTV format or several 4:3 aspect ratio standard-quality picture formats? Some observers in the field think that many broadcasters will do the latter because many more advertising slots will become available.

Also, what can the public expect in the form of a cable terminal with the capability to handle mixed signal formats? The SCTE and Cable Labs have done a lot of work in trying to resolve the specification of the best cable modem. Truly, this device is an important component needed to put the cable television industry into the next millennium. Many consumers still are concerned that their current television equipment will become obsolete and ask if some converter will be available to convert the digitally transmitted signals to NTSC format.

The Advanced Television Standards Committee (ATSC) of the FCC has recommended the use of the 8-VSB digital signal transmissions standard. The cable television operators, however, feel that QAM-64 or QAM-256 is a better method and will promote its use on cable systems. TV sets somehow will have to handle both signal types, and the IC chip manufacturers will respond by providing the necessary components.

5.41 NTSC and Digital Techniques

Digital television techniques have been around for nearly 20-odd years in the form of time-base correctors and video frame store devices. Most cable and broadcast television studios regarded such devices as simply black boxes because analog signals went in and came out. The digital technology simply operated inside the box. VCRs were also subsequently developed that used digital circuitry. The Society of Motion Picture and Television Engineers (SMPTE) then responded with standards for an all-digital television production studio. Thus, the frame store's switching and recording functions were performed in the digital domain.

5.411 Digital technology was developed to operate with the standard NTSC video format in use today. The video signal is converted to a digital signal by an analog-to-digital converter (ADC) by conventional means in which the resulting digital bit stream is processed and then converted back to an analog signal by a digital-to-analog converter. The processing might simply be a signal delay or frame store function. Digital parameters, such as sampling rate or a number of bits per sample that work with a 4.2-MHz bandwidth video signal, are handled within the device and are not under control of the user; hence, the phrase "black box" is used.

To convert the NTSC color television signal voltage to a digital data stream of pulses, an appropriate ADC is used. A basic ADC is shown in Figure 5-26a. In order to convert back to an analog signal the reverse process has to take place; specifically, a digital-to-analog converter (DAC) has to be used. The block diagram for a DAC is shown in Figure 5-26b.

To convert analog signals to digital signals, certain rules dictate the sampling rate and the number of bits per sample. When transmitting digital data through a communication channel, the necessary bandwidth and the signal-to-noise ratio control the accuracy and quality of the digital signal. Resulting errors in the binary bits are referred to as the bit error rate (BER). This is an important measurement of digital signal quality. The input analog signal frequency is an important specification that controls the sampling rate and conversion to a digital data stream.

After much research and testing for the video signal, it was determined that the sampling rate should ideally be four times the chroma subcarrier frequency, or 14.3 MHz. For excellent signal resolution, a choice of 10 bits per sample results in a serial bit stream of 10×14.3, which equals 143 Mbps. The minimum bandwidth needed to pass this signal is equal to one-half this, or approximately 72 MHz. Clearly, this signal cannot be

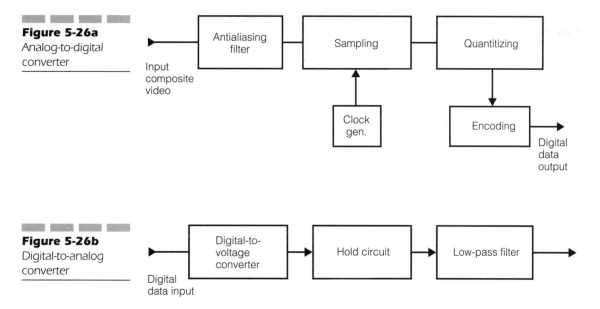

Figure 5-26a
Analog-to-digital converter

Figure 5-26b
Digital-to-analog converter

transmitted in the standard 6 MHz television band without some form of compression. Some of the rules and computations for NTSC video analog-to-digital conversion are shown in Example 5-2. Digital television signals are used in studios and post-production facilities and are not transmitted over the air. Therefore, the necessary bandwidth requirement is not an issue.

5.412 In this section, the process of forming a television image in the digital domain will be addressed. A system of progressive scanning is used, as opposed to interlaced scanning for analog NTSC television. A superior television picture will result by essentially increasing the number of pixels over the surface of the television screen. With NTSC 4:3, 525 line pictures, the horizontal resolution is a function of video bandwidth and is defined as the number of picture details that appear across three-quarters of the screen width. The number of horizontal details can be computed by multiplying the bandwidth in hertz by the horizontal scan time, giving 449 details. Three-quarters of this number gives 337 lines of horizontal resolution.

For vertical resolution, the total visible number of horizontal lines multiplied by 0.7, a utilization factor (Kell factor), gives 338, which is almost the same as the horizontal resolution. The pixel number corresponding to this vertical and horizontal resolution is simply the product of 449 ×

Example 5-2

According to Nyquist's sampling theorem, the sampling rate or frequency

$$f_s = 2f_b$$

where f_b = upper frequency limit or bandwidth. For $4 \times f_{sc}$ where f_{sc} = 3.58 MHz,

$$4 \times 3.58 = 14.3 \text{ MHz}$$

For 10 bits/sample,

$$\text{total bit rate} = 10 \times 14.3 \text{ MHz} = 143 \text{ Mbps}$$

Since $f_s = 2f_b$, $f_b = (f_s)/2$, and BW $= f_b$,

$$\text{BW} = \frac{143}{2}$$

$$= 72 \text{ MHz}$$

338 for 151,762 total pixels in a picture frame. This calculation is a good milestone and useful in comparing NTSC quality to other picture-making methods.

For color video transmission, the composite high-digitizing sampling rate at four times the chroma subcarrier frequency (4 fsc) can be used. With NTSC, the digital signal can be broken down to the Y or luminance video signal sampled at 4 fsc, and the I and Q chroma signal sampled at 2 fsc. When the Y signal and the chroma difference signals are digitized, Y is sampled at 4 fsc and the B-Y and R-Y at 2 fsc. This method is known as 4.2.2 sampling. The actual sampling is synchronized with the video signal chroma subcarrier, which is related to the horizontal scanning rate, thus establishing the sampling points on the video waveform.

The important point is that the greater number of picture-making elements of proper luminance and color, the better the picture quality. This allows for larger screen sizes in which any picture distortions are amplified. Each picture consists of a frame of pixels over the face of the screen with the position of each pixel fixed in location. The pixel locations on the screen form a matrix of vertical and horizontal pixels. The relationship between horizontal and vertical pixels, aspect ratio, the frame rate, and bit rate is shown in Table 5-4. Of course, digital compression has to

Table 5-4

Comparison of
Uncompressed
HDTV and NTSC-
Quality Digital
Pictures

	H Pixel	V Pixel	Aspect Ratio	Frame/ Field Rate	No. of Bits	Bit Rate (in Mbps)
HDTV	1920	1080	16:9	60	16	1990
NTSC-Quality	640	480	4:3	30	16	147

be done so that digitally encoded television picture signals can fit into a realizable communication channel.

5.413 The compression method chosen for broadcast television is the MPEG-2 standard developed by SMPTE that compresses the high-bit rate digital television signal into a 19.5-Mbps compressed rate. This compressed rate is now transmitted as a 6-MHz bandwidth signal using 8-VSB modulation. The purpose of video or digital compression is to reduce the transmission bit rate to a value compatible to the transmission medium. Because NTSC video quality with a 4:3 aspect ratio requires approximately 150 Mbps to transmit pictures, a large value of bandwidth from the transmission medium will be required.

It is a well-known fact that successive pictures contain a large amount of redundant data. For example, an outdoor scene will have a nearly cloudless blue sky from picture to picture; thus, this same information is relatively constant from scene to scene. Many methods are used to reduce redundant information. The motion picture industry has been a big help in analyzing picture scenes and the way we see and perceive images.

Compression methods have been explored and studied in great detail. The MPEG-2 method produces excellent results and is the method recommended by the Advisory Committee Advanced Television Service (ACATS). One of the earliest compression standards was known as MPEG-1 and was used as a digital storing method for video and audio programming at 1.8 Mbps. This method produces a picture and audio quality similar to that produced by VHS cassette recording. MPEG-1 can compress 100 Mbps (4095 × 4095 pixels) to a 1.8 Mbps bit rate, a 67:1 compression ratio.

MPEG-2 is an extension of MPEG-1. Any MPEG-2 decoder can handle MPEG-1 data. Thus, some compatibility between the two methods is maintained. As might be suspected, the MPEG-2 compression method is quite elegant and produces a constant bit rate (CBR). MPEG-2 can adjust to fast-moving picture components by using motion vectors that point in the direction of pixel movement.

Picture frames facilitate storing and switching functions. A bandwidth of 6 MHz as specified by the FCC for television transmission requires that compressed 19.4 Mbps MPEG-2 signal will have to use some special modulation techniques. The 8-VSB is the accepted method for broadcast television. Cable systems prefer and will most likely use QAM-64 or QAM-256 for their modulation method. Whatever method is used, receivers for the public and cable subscribers will have to be able to detect both modulation methods. Some industry gurus are suggesting that most of the channel selection, detection, and decoding will take place within the cable modem, and the screen will handle and present the picture and audio signals.

Audio in the MPEG-2 format is also compressed. To cover the musical range of 40 Hz to 16 KHz, a sampling rate of 48 KHz is used, which is greater than the Nyquist minimum of 40 KHz. At 16-bit resolution, a stereo (two channels left and right) will produce a bit rate of 1.54 Mbps. For a surround sound, or multitelevision (MTS), a 4.5-Mbps rate is necessary.

Computer-type audio compression is used in producing CD-ROM and is also used by the direct-broadcast satellites (DBS) systems. These methods produce a properly compressed audio signal that can fit into MPEG-2 data frames. The audio and video data signals are formed into data packets that can make up the program data stream. The digital data packets making up the MPEG-2 compression for both video and audio program content are shown in Figure 5-27.

The data stream for the MPEG-2 structure is simplistic in that it represents the basic format. There can be two audio packet formats consisting of mono audio or MTSC sound. Essentially, the various packets shown are sub-packetized with header information containing information about the data contained in the packet payload. The cable technicians, from a practical standpoint, do not need to be excessively concerned about the format, except to have a general knowledge of the digital data stream.

Figure 5-27
MPEG-2 program bit stream

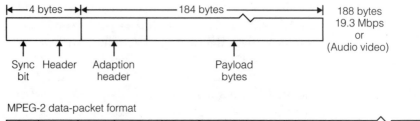

Data packets make up the MPEG-2 transport bit stream.

It is this data stream at 19.5 Mbps that will be modulated in 8-VSB by the television broadcasting industry and transmitted on cable by QAM-64. The question remains, Will the cable television industry receive off-air 8-VSB signals and carry them on the cable in the same format or will they demodulate them and re-transmit in QAM-64 format? Early signs indicate that if consecutive channel 8-VSB signals exhibit any interchannel interference, then it may be necessary to detect and retransmit in QAM-64 format. Ideally, cable operators should work closely with the broadcast industry to receive the MPEG-2 data stream of the broadcast stations signal via fiber-optic connection that will preserve signal integrity. Today, the cable operator can carry the stations' signals in QAM-64 format.

5.42 Digital Television Transmission

At the time of this writing, the digital TV set is being introduced to the public. Because the transmission system is still basically RF for television broadcasting, the digital television receiver will need the tuner to select the appropriate carrier frequency band. The tuner will produce an IF frequency containing the digital modulation. Detecting this frequency will produce the MPEG-2 bit stream. An integrated circuit chip set has been developed to process the MPEG-2 bit stream into video and audio signals. Because it is expected that the digital television set will be able to process many digital television formats, several circuit chips will be needed to do so.

5.421 Digital television as broadcast in MPEG-2, 8-VSB, or QAM-64/256 requires a TV set with digital-processing circuitry to provide the picture and multichannel television sound. The cost, as projected by the consumer electronics industry, is regarded by many as excessive. Because digital television transmission renders the NTSC TV set useless, it has been proposed that some type of converter should be made available to make the conversion. Undoubtedly, this converter will be extremely complex and hence expensive. At this time no manufacturer has said they will have one available.

We have a market waiting for a product. It is clear that this converter will have to tune (select) the channel, demodulate the signal to recover the MPEG-2 and bit stream, and decode the bit stream—functions that will also have to be performed by a digital TV set. This converter will have to take the video and audio data and process it to develop an NTSC video and audio format. This is no small or inexpensive task. As the saying goes,

"Blessed are the chip makers for they will develop a chip," because the market exists for this converter. The development of such a converter has been mentioned in the technical literature from time to time and many think it will be developed.

5.422 The digital television receiver specification has been developed by the ATSC group of the FCC. As it is for most receiver transmitter links, the receiver follows the reverse process of the transmitter. Because the demodulated signal is a digital bit stream, the bit stream has to lock bit synchronization so the packet's video and audio can be found. The transmission layer has to be decoded as well as the transport layer.

The security control module or conditional access module contains the instructions for the receiver's de-encryption process. The transport layer provides the transmitted audio, video, and data packets, which are demultiplexed into their respective digital formats and sent to the video and audio processing circuits. Computer information consisting of games and computer displays are inputted through the IEEE P1394 I/O port. The video format converter drives the video display with the information needed to form the images. A generalized block diagram is shown in Figure 5-28.

Repair and maintenance of digital television receivers will certainly be far more complicated than NTSC-type sets. Highly trained service technicians using specialized test equipment will be required to service and maintain digital television receivers. Many hope that HDTV receivers will

Figure 5-28
Digital television
receiver

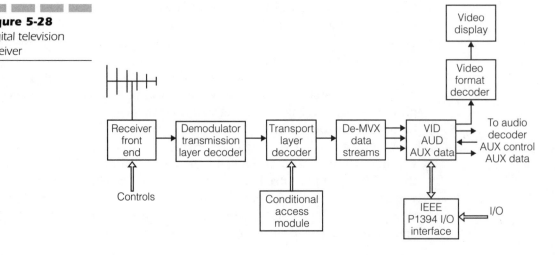

be robust and require little if any service. This will most likely be true if the large-screen flat displays are developed that do not use high voltages and generate heat.

5.423   The modulation method used to broadcast digital HDTV signals is the 8-VSB format. The block diagram describing this method is shown in Figure 5-29. From studying the figure, it is evident that this type of transmission is very different from what we are used to with the standard NTSC method. Digital signal modulation of an RF carrier takes place in many forms. Early studies of 8-VSB transmission have suggested that there is an adjacent channel interference problem possible, depending on the transmission method of the adjacent channel. At present many cable operators are anxiously awaiting the field trials before making any decisions, but the cable industry feels that QAM-64 would be the modulation method of choice for digital transmission through the cable system.

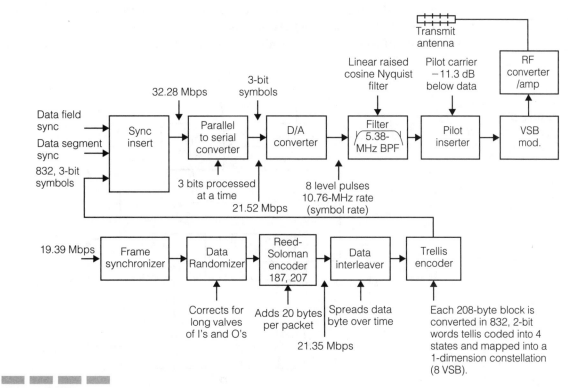

Figure 5-29
A block diagram of an 8-VSB transmitter

5.424 QAM-64 modulation resulted from early digital transmission methods used in computer telephone modem development. This development resulted in higher modem equivalent bit rates such as 14.4 Kbps to 28.8 Kbps, which should be familiar to many of us. QAM method of modulation uses both phase and amplitude modulation to achieve high bit-rate transmission.

As an example, a short study of QAM-16 will demonstrate the method QAM. A block diagram of a QAM-16 modulator is shown in Figure 5-30. The incoming bit rate is divided by two into I and Q channels. Two bits each are fed to the two, four-level PAM converters in each of the I and Q channels, where they are AM-modulated to an in-phase ("I") or to 90° shifted ("Q") carriers. The output signal is formed by summing the output of the I and Q channels. The constellation diagram for QAM-16 is shown in Figure 5-31.

This example demonstrates the principles of operation for QAM digital modulation. QAM-64, which has been selected by the cable television industry as appropriate for transmission through amplifier cascades of a coaxial cable system, has more phases and amplitudes and operates similarly to QAM-16. QAM-64 has a dense constellation diagram and is more difficult to analyze. Cable systems will require QAM-64 modulation

Figure 5-30

Block diagram of a QAM-16 modulator

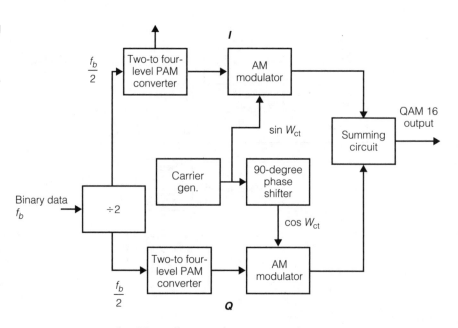

f_b = Bit rate (frequency)

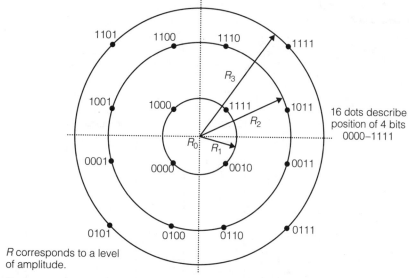

Figure 5-31

QAM-16 constellation diagram

16 dots describe position of 4 bits 0000–1111

R corresponds to a level of amplitude.

This diagram can be presented on an X-Y oscilloscope by demodulating the I and Q signals, where the I channel is connected to the horizontal (X axis) and the Q channel is connected to the vertical (Y axis). The oscilloscope's Z, or intensity, axis should be activated during each bit position.

The oscilloscope constellation diagram should be exactly symmetrical, with small clear dots identifying the bits. Smearing dots indicate noise and nonsymmetry phrase errors.

equipment installed in their headends to transmit the digital television signals.

5.425 The headend for cable systems in the upcoming years is going to have to be much larger and more complicated, requiring more maintenance, testing, and record keeping. The addition of digital services as well as analog services will exist side-by-side until the phaseout of analog services takes place.

Already many cable systems have added fiber-optic equipment in the headends that transports the combined signal to many nodes deep into the service area. Some of the larger and more progressive cable systems have activated the reverse or upstream signal path back to the headend. The network topology for the reverse system is often referred to as the reverse tree architecture because the forward system is known as tree and branch architecture. To solve the problem of reverse system noise combining at the headend, upstream fiber-optic methods are used to divide the noise-source areas of the distribution nodes.

Figure 5-32

Headend
configuration

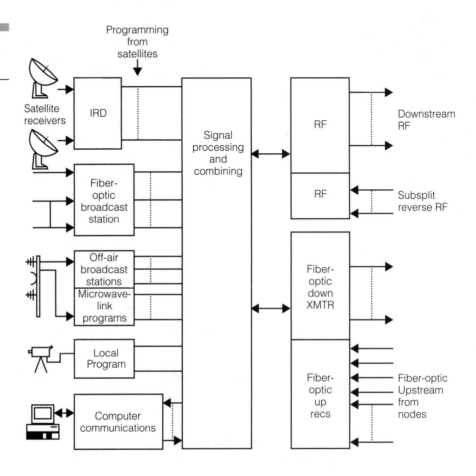

The headend as it exists today is the station where all signals exist. From there, upstream data signals are routed to their proper destinations. Signals from the various sources are routed to the various distribution nodes where they are distributed via short coaxial cable systems. A variety of test equipment and monitoring procedures are required to keep today's system headend in proper operational condition. A block diagram of a type of headend configuration is shown in Figure 5-32.

Of course many types of headend configurations can be found in systems today. Because reliability is now of utmost importance in cable systems, uninterruptible power supplies, signal monitoring, and alarms as well as a stock of spare equipment are all necessary. Above all, the level of training and expertise of a system's technical staff is of utmost importance.

5.43 Digital Data on Cable Systems

Many cable operators know well that providing digital data communications can be a lucrative endeavor. Improvements have been made by cable operators to offer a variety of enhanced services to its subscribers. Such services include the insertion of a comprehensive program guide of locally generated advertisements and providing some forms of pay-per-view services. Many cable operators have redesigned and rebuilt the return signal path of their systems that make high-speed bidirectional digital communications possible. Now the capability exists to offer telephone services over the cable network. This is called voice over the Internet protocol, or VOIP. Such highly developed cable networks now need to be managed and monitored to keep a high level of quality of service (QOS). Standards such as data over cable service interface specifications (DOCSIS), which describes the necessary protocols, have been instituted for cable networks that provide digital services. The set-top subscriber box can communicate with the network to order television programming and have it charged to the subscriber's account. Digital set-top boxes are available with extended memory that has the capability of storing programs. Now video on demand (VOD) can be downloaded and viewed at the convenience of the subscriber. The set-top gateway (DSG) uses DOCSIS protocol for messaging.

5.431 VOIP appears to be a workable method to providing telephone service over the cable system. Subscribers with a cable modem providing high-speed Internet access can communicate with other computers on the Internet with voice and video information. Although the voice quality is reasonably good and the pictures (on the computer screen) are good as well, VOIP still does not offer the equivalent of telephone service. Latency can cause annoying lapses, but if the cable network is operating properly, this most likely will not become a problem.

Telephone service through the local telephone networks offer what is known as "five nines reliability" (99.999 percent). This means the network is up and active 99.999 percent of the time—this translates to approximately 5 minutes of outage time per year.

For many cable systems, this kind of reliability is not attainable. Also, voice intelligibility has to be very high. This means that the voice frequency band has to be adequate. Latency has to be held at a minimum. Some cable operators are requiring subscribers wishing such VOIP service to install standby power for their terminals. The VOIP service box will provide a telephone jack for a telephone instrument. Many cable operators are planning to provide electric power through the service drop. They see that if they can provide equal or better service quality at near or lower prices, VOIP service could become profitable.

5.432 Cable operators feeling the pressures of competition from satellite systems look to VOD services using set-top devices with memory as a way of providing subscribers with equivalent programming. Some of the drawbacks of satellite services are maintenance of subscriber equipment and the occurrence of sun outages twice a year. Also it is prohibited for some locations because of local restrictions or requirements. Offering essentially the same programming makes the cable system a one-stop provider.

5.433 Internet service is available by many cable operators. Such protocol uses a fast Internet engine, for example, "Road Runner." Such services require a high-speed up and down path for sending files and or receiving data files. Because the return path is also terminated at the headend, a cable modem termination system (CMTS) has to be in place. This device uses DOCSIS protocol and provides the necessary bidirectional communications to the subscriber's cable modem. The speed of communication varies with the number of subscribers using cable modems. In general, cable modems provide the fastest round-trip data communications. The telephone companies offer digital subscriber line (DSL) service as well as asymmetric digital subscriber line (ADSL) service as competition to cable. However, the cable system uses optical fiber as well as coaxial cable, the latter having a wide-signal bandwidth. Telephone systems also use optical fiber in their networks but then use unshielded twisted pair (UTP) copper wires to the subscriber, which limits high-speed operation.

Summary

The ability to convert analog quantities as described by an electrical signal to a digital quantity electrical counterpart is important. This procedure is referred to as analog-to-digital (ADC) and digital-to-analog (DAC). These conversions are readily performed by integrated circuits (chip technology). Coding and decoding of digital quantities as well as storing and recalling these digital quantities (numbers) are significant features in our computerized world.

Communications in the digital domain developed into LANs, where digital data was distributed to various terminals and workstations. Storage of digital data files is the work of servers that act as libraries for digital data, where data can be requested by workstations and where work can be stored. The communication rules or protocols for the network data transfers have standards, Ethernet being one of the most common. LANs often are expanded into wide-area networks spread over many miles and terminals.

All cables have signal loss, but metallic cables have the most loss. Coaxial cable has less attenuation of signal level loss than twisted pair copper of the telephone type, and therefore the wider signal bandwidth allows for high-speed digital data signals. The network difference between the telephone standard network and LAN technology is that the telephone system is a circuit-switched type, and a LAN is a digital packet–switched network. Both have good and bad points.

The telephone system, because it was in operation long before computers, was used as the first option for connecting computers together. Early modems provided the interface between a computer and the telephone system and were quickly developed for the business world. Because twisted pair telephone cable offered high signal attenuation and a narrow frequency bandwidth, data speed was slow. As modems developed and chip technology improved, telephone modems allowed higher data speeds greater than the frequency band cutoff. This seemed like magic.

Fiber-optical technology offers high bandwidth high-speed data transmission over longer distances than metallic cables. The telephone company developed a method and equipment called SONET. SONET networks carry telephone digital signals at high speed across the continent. The development of digital television and high-definition television called for new methods, equipment, and standards. Digital television signals have to be transmitted at relatively high speeds and call for high-speed wideband techniques. Many cable operators have installed fiber-optical trunks in preparation for digital HDTV service. The broadcast television stations are going to be required to transmit (wireless) television signals in a format called eight-level vestigial sideband type modulation (8-VSB). Because cable operators carry television program signals on consecutive channels, another digital format is necessary. This format is known as quadrature amplitude modulation (QAM) and occurs in several levels. QAM-64 and QAM-256 are the two most used by cable operators.

Digital television service has several features that are attractive to subscribers. One is video on demand (VOD), where with a push of the subscriber remote control, a program may be ordered and delivered quickly. Two-way communication (digital) has to be maintained between the head-end where the video servers reside and the subscriber's digital set-top box. Cable service technicians have to be computer literate as well as trained in digital technology and testing procedures.

Questions

1. Give an example of a device that uses an analog-to-digital converter (ADC) circuit.

2. Name the current term given to describe the resolution of a display screen.

3. Name three types of network topologies.

4. An Ethernet LAN requires data to be placed in what kind of format?

5. What are some of the principal reasons that make the use of optical fibers in a communications network attractive?

6. Name some of the reasons that have allowed the telephone system to last so long.

7. What type of basic network topology does the telephone system use at the local level?

8. Describe the basic formation of the telephone company T-1 carrier.

9. What is the process called that combines T-1 carriers transmitted through the telephone system?

10. What is the slowest data rate transmitted by the synchronous optical network (SONET)?

11. Explain the purpose of compressing the digital video signal.

12. What is the difference between QAM-16 and QAM-64?

Problems

1. Refer to Figure 5-3. For the voltage value of 5 volts, find the corresponding 4-bit digital number.

2. Refer to Figure 5-6. Write the ASCII code for the word *data*.

3. For the telephone systems T-1 carrier method as described in Example 5-1, calculate the time for one binary bit.

4. Prove that an NTSC television video signal that has been properly digitized into 10-bit digital numbers cannot be passed through a channel with a 50-MHz bandwidth. Hint: consult Example 5-2.

Subscriber Installation and Terminal Devices

Objectives

After learning the material in this chapter, the student will be able to

- Understand the importance of the subscriber service drop.
- Describe both the aerial and underground drop installation procedures.
- Explain the use of signal splitting and combining devices used in the service installation.
- Describe the grounding bonding procedure to protect the service drop from the effects of lightning.
- Explain the effects of stress and aging of the service drop and how it affects the signal and the grounding and bonding.
- Describe the development of the subscriber converters or set-top boxes.
- Understand the signal security and theft-of-service problem faced by cable operators.
- Describe signal scrambling and the difference between it and signal encrypting.
- Explain the basic operation of the digital set-top device.
- Explain the two-way digital control between the headend and the subscriber terminal.

6.1 The Subscriber Drop

The subscriber drop, unfortunately, has not had the attention and consideration it deserves from many cable operators. The nature of the cable television industry is partly responsible for poor and improper subscriber drop installations. At system turn-on or activation, the marketing and sales effort produces many subscribers ready to be installed. Most systems do not have enough technical personnel to install the initial influx of subscribers, making it necessary to engage the services of outside drop-installation contractors. Traditionally, drop-installation contractors go from system to system installing subscribers for various cable system operators. Unfortunately, due to the lack of supervision and quality control, poor installations often occurred. In many cases, the discovery of drop problems occurs long after the drop installation contractors have departed. Many systems find that a lot of corrective work has to be done to the drop installs to bring them up to proper specification. In many cases, this corrective work is unforeseen and unbudgeted, causing unplanned expenses.

6.11 Early Installation Techniques

The aging of the drop install has been a problem for many cable operators. The improvement in the design of the "F"-type connector has undergone several major changes. The latest type, as mentioned in Chapter 1, is of high quality and with proper installation should last a lot longer than the first-generation connectors. Often, if subscribers have not complained of drop problems or no service calls have been performed, the original installation will be in operation even to this day. Systems expanding subscriber services by activating the upstream signal path quickly discovered serious noise and interference problems, which is one of the main contributors in the subscriber drop. Many subscribers move the outlet from room to room, adding more cable and connectors to the internal system. The quality of components as well as the subscriber's work is usually substandard and contributes to return system noise. For many systems, the upstream frequency band is 5 to 40 MHz and is referred to as the subsplit return band. Noise generated in the subscriber drops is coupled into the upstream reverse amplifier cascade. Noise and interference is coupled at feeder-splitting points back into the bridging amplifier reverse module. Combining again takes place at trunk-splitting points and is transmitted through the return amplifiers to the headend. An example is described in Figure 6-1.

Since the early days, drop cable has undergone many improvements. Probably the most significant improvement has been in the increased shielding effectiveness. Present-day high-quality drop cable has an aluminum foil wrapping the polyethylene foam dielectric and is covered with two woven aluminum braids, followed by the polyethylene jacket. This type of cable gives a 90-percent shielding that amounts to over 100 dB of isolation. Essentially, this type of cable has three shields and is often called tri-shielded cable. Systems that experience significant noise buildup will have to budget a work project to prepare the return band for upstream signal transmission. A measurement of noise in the upstream band should be made at the headend at points arriving from each trunk branch. Once a particularly noisy branch is identified, testing back along

Figure 6-1
Upstream noise
combining

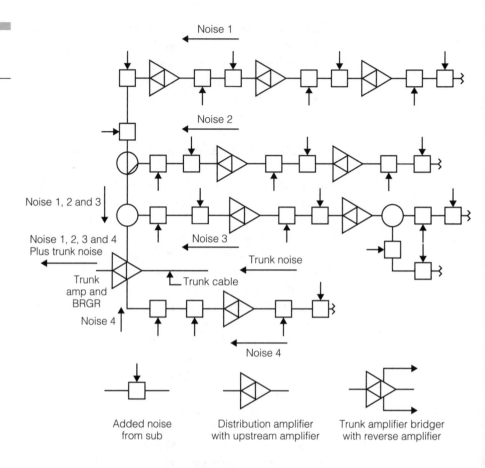

the trunk to the splits should be done to identify which, if any, is the noisiest point. Many systems have found that reworking the drops in the system using 90-percent shielded drop cable, new "F" connectors, and exterior drop hardware drastically improves the noise and ingress on the return path. Troubleshooting maintenance techniques in testing is covered in more detail in the next chapter.

6.111 Subscriber installations in many communities of single-family homes or duplexes are aerial drops. The aerial drop, exposed to the outside weather, will deteriorate quicker than many underground drops. This is mainly due to wind, rain, ice, and snow stressing the drop cable at the points of connection, that is, at both the utility pole and the house. An aerial subscriber drop is shown in Figure 6-2.

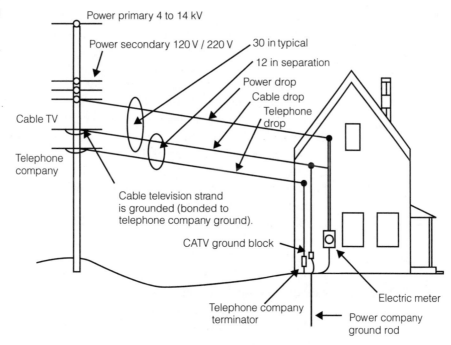

Figure 6-2

Subscriber cable drop

The effects of lightning can also cause drop problems. If a lightning strike is close by, a power surge can take place, or the cable grounding point can be affected to the point of burning or scorching the connector at the entrance point. If a signal outage occurs, a subscriber will request a service call and the cable technician will properly repair the problem. The effects of wind and weight of ice can cause the cable to lose its concentricity, resulting in the increase in return loss and the shielding effectiveness of the cable. Systems experiencing such drop-cable problems usually change to messenger cable, which gives added strength and support to the drop.

The material used in the subscriber drop was similar to that used by telephone companies, but it did not take the cable industry long to develop hardware and drop components that were more appropriate to its own requirements. The ease of installation and high-quality materials are the criteria many progressive cable operators used regarding their subscriber drops. However, many cable operators have elected to use low-cost materials and poor drop-installation contractors, which result in poor service to subscribers and a flood of service calls.

Because the points of connection at the utility pole and the subscriber's home are where the stresses appear, the hanging devices are extremely important. At the utility pole line, a span clamp is used to clamp to the steel messenger strand and provide a captive hook to fasten the drop wire. The drop cable containing its own steel messenger wire can be used by simply wrapping the stripped back steel strand tightly with several turns around the span clamp hook. For RG-6 cable that is non-messengered, a preformed galvanized wire forming a loop with the ends tightly twisted can be used as a cable grip. The twisted ends are coated with a sand-encrusted, semisoft plastic covering. This device is referred to as a pre-form, which in its larger sizes is used in pole-line construction, connecting strands together with splices or dead-end loops. The small version used in subscriber drops is the best method where non-messengered drop cable is used.

At the subscriber's house, a screw hook is used to fasten the drop cable. In many instances, installers fail to place this hook firmly into solid wood, causing the hook to pull loose and the drop to fall. This hook provides proper length to pass through the trim board into a corner timber. A twist in the hook eye prevents a cable using a preform to swing out of the hook. At this hook the cable is looped out and either stapled down the trim board or fastened with cable clamps down the side of the house to a grounded entrance block. This block contains a grounding connection for the ground wire connected to a grounding electrode. The grounding electrode is the electric utility ground rod or connecting wire to this rod. Cable operators should never loosen any clamps connecting wires to the ground point or rod. The proper technique is to add another ground clamp to the rod for the cable ground wire to attach. Drop connections for aerial drops are shown in Figure 6-3a and 6-3b for both the system point of contact and the subscriber's home. Proper hanging and placement of the drop wire is directly related to the life of the installation. The drop wire should not be drum tight or hang too low. A good rule of thumb is to allow a sag at the midpoint of 1 ft per 100 ft of drop cable at a temperature of 70° F. This is essentially eye-balled according to how the drop wire sags over the drop length. Most competent installers can estimate the sag by how hard it is to pull the drop tight.

Drop contractors and cable company installers often work as a single-person crew. Single-person installers often start in the house by placing the cable supply reel outside, drilling the access hole into a basement area or crawl space beneath the house, and threading the cable to the terminal location. At the terminal location, a connector is attached and the cable is fastened to the house timbers by either staples or cable clips. Staples

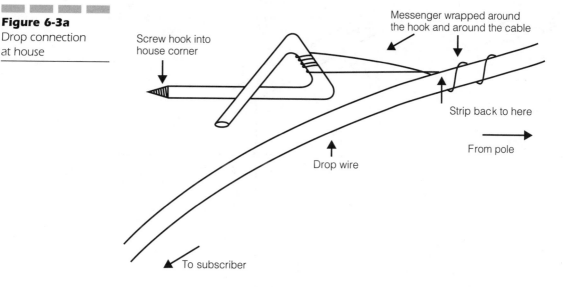

Figure 6-3a
Drop connection
at house

Screw hook into
house corner

Messenger wrapped around
the hook and around the cable

Strip back to here

From pole

Drop wire

To subscriber

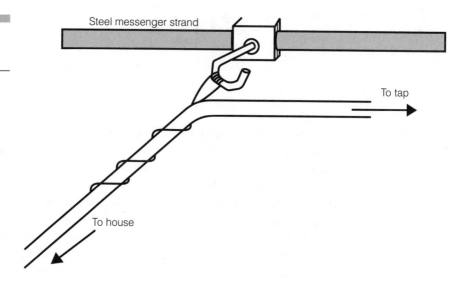

Figure 6-3b
Drop connection
at tap

Steel messenger strand

To tap

To house

should be either Monel or stainless steel composition. Cable clips are usually plastic fastened by aluminum or stainless steel nails. Once the initial inside work is finished, the installer can roll the excess cable back on the reel. The installer should now place the ground block near the hole where the cable comes through, install a connector on the cable, and connect to one terminal on the ground block, making a loop as shown in Figure 6-4.

Figure 6-4
Detail of aerial
installation

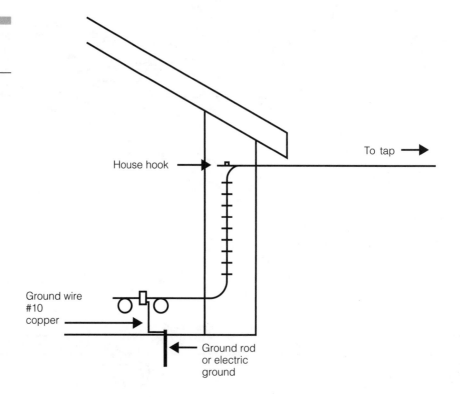

Now the messengered drop cable is used for the actual hanging wire. The messenger wire should be separated from the cable a length approximately equal to the distance from the ground block to the house hook. Now a connector is installed and connected to the ground block through a similar loop and stapled or fastened to the side of the house with cable clips. The messenger wire is formed around the house hook and back around the cable, previously shown in Figure 6-3a. At this time, the installer moves the cable supply reel to the pole, climbs the pole, and cuts a portion of the messenger loose, separating enough to wrap around the span clamp placed on the strand convenient to the tap port. Tensioning of the drop is done at this time by pulling the messenger wire through the hook and wrapping it around the hook and the drop wire, as previously shown in Figure 6-3b. A connector placed on the drop wire and connected to the tap completes the pole work.

The installer should ground the ground block to the power-ground electrode, if this step has not already been completed. If no ground is readily available, a cold-water pipe exposed as a hose connection can be used, providing there is a direct connection to a metallic municipal water supply. If a separate ground rod is required, a 5/8-inch copper-clad rod is preferable to other types. It is often helpful if the cable operator's technical department consults the municipal wiring inspection department about any local codes and grounding and bonding practices.

6.112   Many cable systems today are opting for underground or buried plants, following the trend of many telephone companies. Installing plant underground minimizes the problems of traffic hazards as well as weather hazards. In many urban residential subdivisions, it is required that all utility services be placed underground.

Underground drop wires have been in existence for a long time. Early underground drops were placed in a slit in the ground made by a flat spade or lawn edger. As the need appeared for faster and deeper drop installations, a variety of powered machines appeared. Present-day machines plow drop cables to depths of a foot at rapid rates approximately 15 ft per min through average soil conditions. Soft, wide tires allow these small cable-plowing machines to cross lawns with a minimum of restoration problems. At the pole, the cable is connected to a tap port and is either stapled or fastened with clips to the utility pole. The bottom 8 to 10 ft should be covered with a molding to protect the cable at ground level. A correctly installed drop cable at the pole is shown in Figure 6-5.

Usually the cable is plowed from the base of the pole to the house where the ground block is located. If a sidewalk is located between the pole and the house, a section of steel or PVC schedule 40 conduit should be pushed under the sidewalk. An appropriate amount of cable should be left at the pole to reach the tap port and dressed to the pole. At the house, enough cable should be left to form a drop loop at the ground block. Direct burial drop cable with a flooding compound should be used for such underground drop installations. For present-day cable systems offering service in the 800-MHz range, only RG-6 or larger 75-ohm drop cable should be used. A typical underground drop connection at the subscriber's house is shown in Figure 6-6. Many new home developments are served by underground cable plant in pedestals or vaults, requiring drops to be placed underground as well.

6.113   Most cable subscribers require more than one cable outlet, and many subscribers have a whole interior cable distribution system. Often new homes are prewired using either a loop-through or home-run system.

Figure 6-5
Underground drop
shown at pole

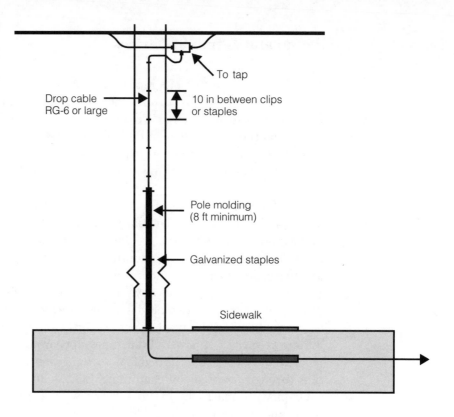

To tap

Drop cable
RG-6 or large

10 in between clips
or staples

Pole molding
(8 ft minimum)

Galvanized staples

Sidewalk

Figure 6-6
Underground drop
to subscriber's house

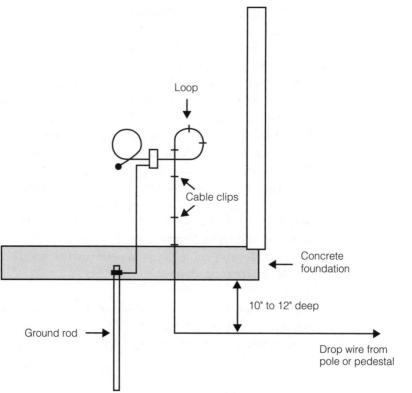

Loop

Cable clips

Concrete
foundation

10" to 12" deep

Ground rod

Drop wire from
pole or pedestal

Both types are shown in Figure 6-7. Multiple outlets require either signal splitters or directional couplers to distribute the input signal level among the outlets. Some cable operators feel that the loop-through system saves cable and hence cost, while other operators think that more directional couplers offset any savings in cable costs. The choices are not often clear-cut and often depend on the expertise of the installer technician. Ideally all subscriber outlets should have nearly the same signal level and enough levels to operate the TV set or converter terminal. Some cable operators

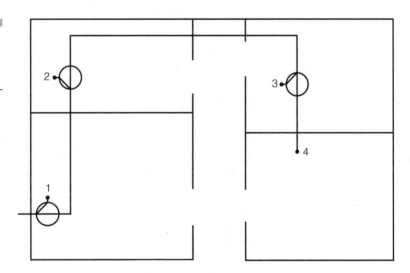

Figure 6-7a
Four-outlet loop-through using directional couplers

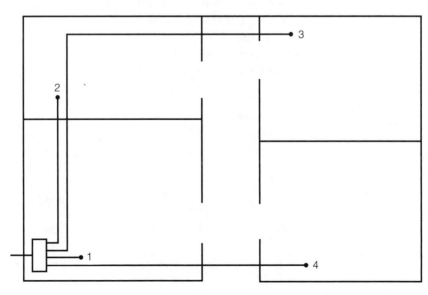

Figure 6-7b
Four outlets of home-run type using one four-way splitter

offer prewire services to developers or subscribers, with costs rebated at activation. Cable operators doing it this way have some control over the quality of the installation. Unfortunately, many subscribers elect to extend or rework their interior cable distribution system using inferior components that are often improperly installed. This results in poor service, for which the cable company is blamed. Systems that have a large number of poor subscriber installations can experience noise and ingress problems in the upstream return path.

Most subscribers choose two or three cable outlets to cover a family room, bedroom, and possibly a kitchen/dinette area. Signal-splitting can often be conveniently made outdoors where the drop connects. Many vendors of drop equipment offer two-, three-, and four-outlet outdoor signal splitters with a grounding connection. This type of connection does not need a separate ground block because these devices are fitted with a ground wire connection. A splitter of this type is shown in Figure 6-8.

6.114 Two of the most important aspects of the subscriber installation are grounding and bonding. Proper grounding and bonding protects the drop as well as the subscriber's equipment from the effects of lightning. In lightning-prone areas, soil conditions often do not provide adequate ground conduction necessary for a good earth ground. Often several ground rods screwed end to end are necessary to obtain a decent earth ground. Some suburban residential areas have a metallic water supply system, which can act as a good ground for the telephone electrical and cable system. Electrical utilities are grounded using ground rods to the water system, provided it is metallic. However, the proper ground electrode for cable systems is the electrode used by the utility companies, which is either a rod or water system.

Ground conduction is usually measured using a device called a "meg-

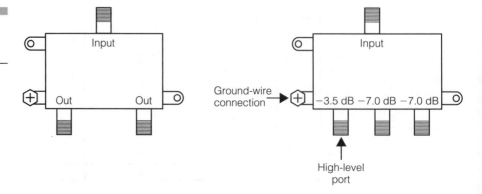

Figure 6-8
Outside grounded
splitter

ger." Two copper ground rods are separated a few feet apart and are driven to the same depth. The instrument is connected to each ground rod, which then places a difference of potential between the ground rods measuring the resulting current. The ground resistance is read from the instrument scale or can be calculated by dividing the test voltage by the measured current. Most instruments of this type provide a direct reading of the ground resistance between the two rods. The proper measurement of ground conduction is the ratio of the ground resistance to the distance between the rods given in ohms per foot. Often the desired result is 25 ohms per foot. If need be, a longer rod or a rod with a larger diameter may be necessary to achieve this result.

The conductor size used to ground the cable service to the grounding electrode is #14 AWG copper wire. This size is the minimum that should be used, and many cable operators use a larger size #12 AWG solid copper conductor with PVC insulation. The insulation offers some added strength to the wire and protection against corrosion. This type is usually readily available from either cable-drop supply vendors or electrical-supply companies. Such suppliers usually carry ground clamps for a cold-water pipe connection or ground rod clamps as well as split bolt connectors that are used to connect to another ground wire. Several types of these devices are shown in Figure 6-9.

The usual installation procedure for proper grounding is to determine how the power neutral is grounded. If the power ground electrode is a driven ground rod, a ground rod clamp can be placed on the rod and a #14 or larger copper wire can connect the cable ground block to the clamp. If the top of the rod is not available, the cable ground wire can be attached to the power ground wire using a split bolt connector or the wire can be connected to the power metallic entrance box using a spade lug and screw.

At the poles, any bonding is done using bare #6 AWG copper wire using strand-bonding clamps connected to the cable system strand and the telephone company strand. The bonding procedures and specifications for an aerial plant are usually spelled out in detail in the pole attachment agreement between the cable operator, the telephone company, and the power company. Proper grounding and bonding is extremely important to the cable operator as well as the utilities. The cable operator's technical staff should instruct all cable installers on proper procedures for grounding and bonding the distribution plant and the subscriber's home.

Several large multiple-system operators (MSO), as they are known in the cable industry, are presently or will soon be offering telephone service. Because telephone systems carry their own power, a loss of electrical power does not cause the telephone system to fail. Cable systems offering

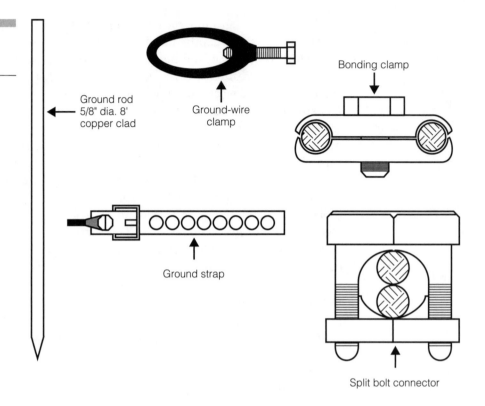

Figure 6-9
Ground system
devices

Ground rod
5/8" dia. 8'
copper clad

Ground-wire
clamp

Bonding clamp

Ground strap

Split bolt connector

telephone service are going to have to provide all signals and power to the telephone terminal in the subscriber's home. This means the drop cable is going to carry electrical power through the install cable path to the telephone set. For proper protection from power surges, a surge protector will have to be placed at the ground block location. This necessitates a cable communication entrance box housing to contain the ground block, surge suppressor, and possibly a signal splitter. Fortunately, several vendors offer a variety of such products to the cable industry. Also, for subscribers not taking any services requiring the upstream path, the placement of a filter in this cable entrance box will prevent any ingress and noise from getting into the return system. The use of an entrance box makes a nice-looking install and keeps the drop equipment and connectors protected from the weather. This adds to the system reliability necessary to compete with other communication providers. An example of an outside residential entrance enclosure is shown in Figure 6-10.

Many new homes, condominiums, and apartment complexes are installing what is known as structured wiring, which means a standard method of entrance and distribution from the entrance terminal to the

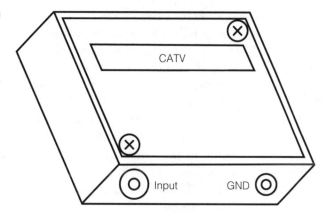

Figure 6-10
Cable communi-
cation entrance
enclosure

subscriber terminal is employed. At present, coaxial cable has a terminal at the area point of entry in the form of the usual pedestal or cabinet, with a connected system ground as well as an installed surge protector. At this area point of entry the telephone networks have their cables connected and grounded. From the outside entrance cabinet the cables run to the interior communications box, which acts as a patch panel connecting the telephones, televisions, and the computers for each subscriber location. Such standards will offer continuity to the installations making it easier to service. Most developers are requiring that service providers have the distribution cables installed in the usual aerial or underground manner and provide a feed to each subscriber's communication panel. Many electrical contractors are expanding their business by prewiring new residential areas in a standard manner using standard equipment. Telephone and cable operators are responsible to connect the telephones and set-top terminals to the subscribers as they move in and order service. Structured communication wiring that falls into the low-voltage category has been promoted by the building industry through Building Industry Consulting Services International (BICSI). This organization has promoted orderly standard wiring practices for intrabuilding wiring providing training through courses and seminars. Technical personnel who take the training program become certified at the level for the completed course.

6.115 Most cable system operators perform few or no quality-control checks on subscriber installations, regardless of whether they were done in-house or by a drop contractor. The reason, most likely, is due to extra expense. This, of course, is a serious mistake that can come back to haunt the cable operator. After all, it is the drop connection that is providing the service the subscriber is paying for.

Two schools of thought exist for checking out subscriber drops. One is to have a technician follow around the installation crews and check on every drop. This is, by far, the most comprehensive as well as the most expensive. The second method is to spot-check the drop installation. Finding a problem with the quality of the work will point the finger to the crew responsible, which is the type of check preferred by most cable operators.

The first thing a cable operator must do to obtain quality drop installations is to make sure that the installers, either in-house or contracted, have proper training in the way the cable operator wants the work to be performed. This includes specifying the drop material and the methods for doing the job specified by the cable operator. Cable operators should start quality-control checks on the installations either on a complete or spot-check basis as soon as the work commences. This will nip any problem areas in the bud.

The cable operator should keep accurate records of which drops were inspected and note any problems and comments. Forms for these inspections can be developed and printed in tablets. The inspector can then check the items listed per drop material and installation as shown on the form. At the end of the day, the inspection form information can either be filed or entered into the technical file computer database. By using a hand-held computer in the field at each inspection location, data pertaining to the inspection can be entered directly. This information can be accumulated on diskette and either filed or compiled into the technical database. The cable industry should, if it not already has, obtain appropriate software to aid in proper record keeping.

Essentially, the information entered on a form or on a computer is the same. An example of an inspection form is shown in Figure 6-11. Whichever method is used, the quality-control information at initial turn-on is important for maintaining the drops, subscriber's service, as well as studying the effects of the drops as the system ages.

6.12 Drop Aging and Deterioration

The aging of the subscriber's drop is definitely related to costs and expenses. High-quality drop installations may cost slightly more in the beginning but will definitely last longer and be relatively problem-free. Unfortunately, most plants and equipment under the conditions of weather and seasonal temperature changes deteriorate with age. Thus, the outside subscriber's drop has essentially a limited life span that varies

Figure 6-11

Installation quality-control checklist

INSTALLATION QUALITY-CONTROL CHECKLIST

OUTSIDE

ADDRESS OF SUBSCRIBER:
    STREET NAME / NO . _____

AERIAL / U.G. _____ LENGTH OF DROP (EST) _____ FT

MIDSPAN / DIRECT _____ DIRECT POLE TO HOUSE (Y/N) _____

GROUND ELECTRODE:
    SEPARATE ROD   Y _____   N _____

ELECTRIC UTILITY :
    GROUND ROD _____ CONDUIT _____
        METER BOX _____   WIRE _____

WATER PIPE CONNECTION _____

GROUND CONNECTIONS:
    TIGHTNESS _____   WIRE PLACEMENT _____

GROUND BLOCK "F" CONNECTORS:
    APPROVED CONNECTORS (Y/N) _____WEATHER BOOTS (Y/N) _____
    USE OF SEALING COMPOUND (Y/N) _____

SIGNAL LEVEL AT GROUND BLOCK:
    LOW CH # _____   HIGH CH # _____

INSIDE

NUMBER OF SPLITS _____   SPLITTERS / COUPLER _____

SIGNAL LEVEL AT FARTHEST POINT FROM GROUND BLOCK:
    LOW CH # _____   HIGH CH # _____

CONVERTER / SUBSCRIBER MODEM:
    TYPE USED _____

SUBSCRIBER'S COMMENTS ON SERVICE AND WORK DONE:
_____
_____
_____
_____

NAME OF TECHNICIAN _____   DATE _____   TIME _____

with the quality of the initial installation. For many systems, the life of a drop installation varies from 8 to 12 years, which is also often the length of the system financial loan. Therefore, the buyer of a 10-year old system should evaluate the condition of the subscriber drops before making any financial offer. A new prospective owner of a 10-year old system should plan to replace the drops as part of an overall system upgrade.

6.121 Metallic corrosion due to moisture is one of the concerns of drop deterioration. Iron rusts, aluminum oxidizes, and copper turns green; all are forms of corrosion. Metallurgists have studied the effects of metal corrosion and the effects of dissimilar metals in contact. Throughout the years, the metal industry has provided corrosion-resistance metal alloys for manufacturers to make metal components for the communications industry. Aluminum alloy amplifier, tap, and passive housings are all used in the outside cable plant. Ground blocks are made of aluminum alloy and are fitted with plated brass cable connections. The latest version of the "F" connector is made of high-quality metals and contains an interior O-ring to keep moisture from entering the connector from the cable end. The manufacturing industry has responded with high-quality drop installation components, and cable systems should use them. Splitters and couplers are made from corrosion-resistance aluminum alloys and some use stainless steel. All are sealed against outside moisture. Still, if some of these enclosures are not hermetically sealed, moisture from the outside air can carry moisture to the inside and water can condense, causing interior corrosion. If such components can survive 10 winters of rain, ice, and snow, this amounts to a 10-year life span for drop components. As subscribers move from one location to another causing subscriber churn, a cable operator should use this disconnect and reconnect cycle to upgrade the drop installations.

6.122 The continual stretching and relaxing of drop cable in the wind causes it to deteriorate. Also, temperature effects of expansion and contraction add to the wear on the drop cable. Since this cable is a transmission line with a 75-ohm characteristic impedance, it must retain its characteristics or the signal it carries can deteriorate. The center conductor must remain in the exact center and an equal distance from the shield, thus maintaining its coaxial condition necessary to maintain its characteristic impedance. If the source (driving) and the load impedance are matched at 75 ohms, the cable has to be 75 ohms over the frequency range in order to maintain the matched condition. When the load or source impedance is different from the cable characteristic impedance, a mismatch occurs and signal reflections take place. These reflections cause the subscriber's pictures to have ghosts, causing serious picture impairment.

Hanging cable usually has prescribed sag, which allows for expansion and contraction. The cable grips mentioned previously, called preforms, keep the hanging stresses from affecting the connector at the tap, which can also be accomplished by using messengered drop cable. Still, the nearly constant motion of aerial cable can cause stress cracks in the outer jacket, allowing water and moisture to enter and start corroding of the cable shielding. Proper attention to the drop during a service call can head off problems later on. A few more minutes spent by a competent service technician observing the condition of the drop installation should be a company rule.

6.123   The effects of lightning on subscriber drops can be catastrophic. Various sections of the United States are more prone to lightning than others, and an isokeraunic map, showing different areas' rates of thunderstorms, is shown in Figure 6-12. The southern and southeastern states are more likely to experience lightning problems. Often these same locations have sandy soil, and maintaining a good earth ground is a problem concerning not only the cable operator but also the electrical utility company and the telephone company as well. As previously discussed, a

Figure 6-12
Annual thunderstorm days

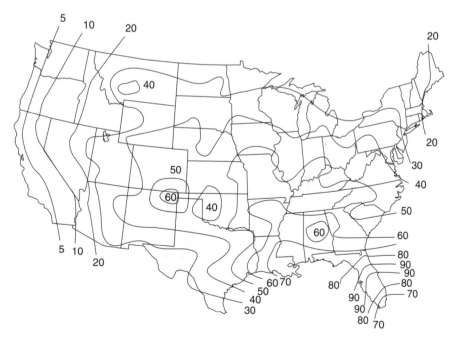

good ground at the ground block is necessary to protect the subscriber's property as well as the cable service.

Cable drops that have been affected by a direct or nearby strike can exhibit scorching around the ground block. If this happens, the ground wire and connections should be checked and possibly the whole ground block and the associated cable connectors should be replaced. All of the subscriber outlets should be tested for signal levels in case some of the components found in the splitters or directional couplers have been burned out. Sometimes the effects of lightning show up at a later date, resulting in a call back. Since the subscriber's TV sets and converters are connected to the power system and the power system ground, proper ground bonding should be done to protect any currents due to lightning to flow in the home wiring. This problem is illustrated in Figure 6-13. The

Figure 6-13
Ground bonding for protection from lightning

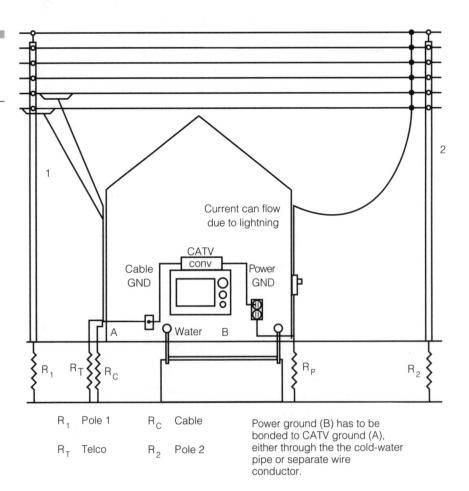

Current can flow due to lightning

CATV conv

Cable GND

Power GND

A   Water   B

R_1 R_T R_C R_P R_2

1

2

R_1	Pole 1	R_C	Cable
R_T	Telco	R_2	Pole 2

Power ground (B) has to be bonded to CATV ground (A), either through the the cold-water pipe or separate wire conductor.

75-ohm to 300-ohm matching transformer required by older TV sets is easily affected by lightning and often has to be changed after particularly heavy lightning. In many cases, these transformers protect the set itself by "taking the hit."

6.2 Subscriber Converters

Subscriber converters have been through many generic versions throughout the years. The first use for a subscriber converter was to enable a TV set to tune to a channel it could not receive otherwise. Early TV sets were basically 12-channel sets, offering channels 2 through 13. When the FCC opened up the UHF television band, TV sets made after that time became fitted with a UHF channel selector, while earlier TV set owners had to purchase a separate UHF converter. As mentioned in Chapter 1, cable operators provided 12 channels on a cable system with an upper frequency limit of 220 MHz. UHF stations were translated down to the band of 108 MHz to 170 MHz, which could carry nine mid-band television carriers. The translation was done at the cable system headend by adding these nine channels to the already established 12 channels, which gave a system 21 total channels. A subscriber converter then translated these channels back up to UHF channels to be selected on the television set's UHF converter.

6.21 Mid-Band Block Converters

This type of UHF converter was considered a mid-band block converter that translated a block of mid-band frequencies up to UHF channels. It used a fixed frequency local oscillator that transformed the mid-band channels up to UHF standard channels. Such a converter is shown in Figure 6-14. A converter of this type had no controls or on/off switches and remained active as long as the plug was in the wall. When pay or premium programming became available, the mid-band channels were used to carry this service. Usually one or two premium channels were grouped together and separated from the UHF channels by a blank channel. This converter also enabled the use of a remote control, but when this type of converter was in use, few TV sets had remote controls.

Figure 6-14
UHF block converter

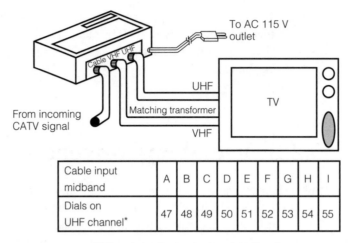

Cable input midband	A	B	C	D	E	F	G	H	I
Dials on UHF channel*	47	48	49	50	51	52	53	54	55

*Other channels may be translated to other UHF channels.

6.211 When premium or pay channels were added to cable systems, some means of preventing the unauthorized use of them was needed. Because a cable system grouped pay channels consecutively, a signal-delete filter (trap) was devised to remove the premium channels from the subscriber's drop. This trap appears in a barrel-type configuration, with a male connector on one end that was screwed into the tap port and the drop connected to the other end. Because this type of trap was used to deny the premium channels from a subscriber, all nontakers of the premium channels had to use this trap.

For systems without many premium program subscribers, a lot of traps were used. This, of course, was an expensive choice. A "fix" to this situation was to insert an interfering carrier into the premium channels that would cause intentionally severe picture distortions. Placing a sharp notch filter in the drop, usually at the subscriber's tap port, would delete the interfering carrier, allowing clear reception of the premium channels. This type of trap was known as a positive trap. Only premium subscribers needed these traps, thus a system with few premium subscribers did not need to use many traps and would save money. If 50 percent of a system's customers were premium subscribers, it would make no difference in costs, regardless of which method was chosen, because both the positive and signal-delete types of traps cost nearly the same amount of money. Figure 6-15 shows what these traps look like.

A system audit of the number of active premium subscribers a system had could easily be performed by viewing the system taps on a drive-

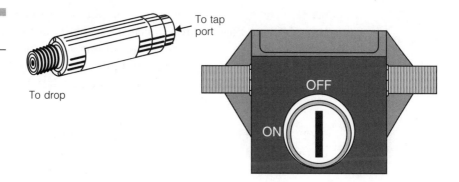

Figure 6-15
Signal-security taps

To tap port

To drop

by basis. Eventually, people stealing the premium service were found by systems that discovered a look-alike dummy trap that did not function. Cable television supply vendors quickly developed a trap security shield that enclosed the trap and required a special tool to either install or remove it.

6.212 The use of converters and traps did have some effect on picture quality. The block converters could drift a bit and thus the subscriber would have to readjust the fine-tuning on the TV set. Leaving the converter constantly powered helped the drift problem. TV sets at the time had poor noise figures for the UHF tuner and because the block converter translated the mid-band channels to the UHF tuner, slight snow could be seen by subscribers viewing these channels. Negative-type traps had to be made with steep edges above and below the pass band so as not to remove any picture information from adjacent channels. Positive traps, while removing the interfering carrier, would often remove some sideband signal components, causing some amount of picture impairment.

6.22 The Selectable Converter

An improvement to this process was the selectable converter, which allowed all program selections to be made by the converter channel selector. The channels selected on the converter were converted to either channel 3 or channel 4 on the VHF channel selector. This selectable converter began many disagreements between the consumer electronic industry and cable operators. Unfortunately, many of these converters were poorly made and caused cable systems a lot of trouble.

6.221 The selectable converter was placed between the cable drop and the television receiver and appeared on top of or near the TV set. Some of these converters also had a fine-tuning control that often confused the subscriber with the television tuning system and the converters. Mistuning problems caused many cable operators a lot of expensive subscriber service calls.

6.222 Signal security with this type of converter was done as before, by using either positive or negative traps as required. Actually, this type of converter did little for the cable operator and in many instances caused more problems than it solved.

6.23 The Programmable Converter

The electronic programmable converter was introduced when signal-scrambling methods were developed. Signal scrambling of the premium pay channels was done at the cable system headend, and it was the job of the converter to unscramble the signals at the subscriber's TV set. The first generation of this type of converter used a "PROM" chip with the decode logic permanently burned into this chip. Cable operators could order from suppliers the preprogrammed chips or could program their own using a PROM programmer, often referred to as a PROM burner. It did not take long for pirate chip operators to appear, allowing unauthorized use of a cable system's premium pay channels. This converter contained the descrambling circuits that would descramble channels authorized by the PROM chip. Again, channel selection was performed by the converter and translated to a low VHF channel, usually channel 3 or channel 4. This made the TV set's remote control useless. Because many subscribers wanted remote control, cable converters with remote controls became available. Now subscribers had two remote controls, and one was useless. The relationship between the consumer electronics industry and the cable system operator as well as the subscriber had not improved.

6.231 Signal-scrambling methodology was in the beginning quite simple and effective. Removing the horizontal synchronization pulse caused TV sets to lose synchronization, destroying the picture, yet the audio signal was not disturbed. In order to remove the horizontal synch pulse, the scrambling circuit had to find it. This is actually quite easy at the circuit level and at each sync pulse interval a circuit shorted it out to ground level (0 v), thus removing its presence from the video signal. Actually, this was done at the video IF inside the scrambler at the headend. In the converter, the reverse procedure was performed on command from the programmed PROM, thus restoring the synch pulse. One of the problems with this converter is they could "walk," meaning it could be moved to another location and be used with any TV set.

6.232 The authorization and security control remained with the PROM. By using the PROM programmer, some cable systems tried to use more complicated authorization codes. Still, for many systems, the signal thieves just became more inventive. Many manufacturers of cable converters worked with cable operators to try to keep up with the theft-of-service problem.

6.24 The Addressable Converter

The addressable-type converter seemed to solve some of the security problems. It had an exclusive code for the subscriber programmed into its PROM. Each converter had to have this code refreshed at predetermined intervals or the converter would shut down and cease operating. Also, the information for the selected channels to decode was transmitted on the downstream data signal along with the address data. This data stream was transmitted just above the FM band using FSK-type modulation. This type of converter allowed a subscriber to call the cable customer office and order a change-in-pay service. The controlling computer, which was also usually a part of the billing system, would select the subscriber code and change the decode instructions for the desired channel and delete any others from the service.

This, of course, was a savings to the cable operator because a technician did not have to physically go to the subscriber's home since the changes would be made in the office.

6.241 Converters shipped from the factory to a cable operator had to be initialized before being placed in service at a subscriber's home. This initialization process could be done quickly if the control computer was fitted with a bar code reader. If not, the converter identification number could be entered on the keyboard. Manufacturers of addressable converters supplied the control software to the cable operators. This software was used to program the controlling computer, which in turn controlled the converters out in the field. During the initialization process, the subscriber's identification code number was assigned and added to the database along with the coded information, as the premium services were also added. The control computer sent converter activation refresh signals along with premium channel selections to the addressable converters on a rotating basis. The more premium subscribers a system had, the longer the refresh time cycle. If a converter was unplugged or disconnected from the cable system, it would automatically shut down and become dormant. This prevented passing the addressable converter around or, as they say, kept the box from "walking." Still, actual signal security depends on the level of signal scrambling or encoding methodology.

6.242 The more complex scrambling is, the more secure the signal becomes. Also, the more complicated the scrambling technique becomes, the more difficult the descrambling method. Of course, the added complexity of the whole process results in a higher cost. The two basic categories for signal scrambling are analog and digital. Analog methods are often regarded as soft and digital methods as hard, referring to the signal-security level.

As discussed previously, signal-trapping methods using either positive or negative traps are regarded as soft signal security. This signal could be defeated simply by removing the negative trap (usually at the tap) or replacing it with a look-alike dummy trap. The positive trap, which was needed to remove the interfering carrier, could be placed at the subscriber's TV set. Such traps are easy to obtain and can often be found in the back merchandising pages of electronic hobby magazines.

A more secure analog method is the sync suppression method, which can be done at RF or on the base-band video signal. The horizontal sync pulses are essentially suppressed a sufficient amount, causing the TV set to lose its horizontal hold. Both methods used a 15.73-KHz sine-wave signal to suppress the sync pulse, which is shown in Figure 6-16. In order to unscramble the signal, the reverse process is used. Essentially, a sine wave at the same frequency is used to restore the synchronizing pulse and the brightness level. This sine wave is transmitted on the cable system either just above the FM broadcast band or on any FM carrier and is transmitted as part of the FM audio signal on the scrambled channel. The unscrambling process is performed in the set-top converter usually just after the input converter circuit. The restored or unscrambled picture is then converted to the set-top output frequency, usually channel 3 or channel 4.

Another analog horizontal sync suppression method is the gated sync suppression method. This method can be used for RF, IF, or base-band video. In the RF/IF methods, pulses with the same timing as the horizon-

Figure 6-16

Sync-suppression methods using sine waves

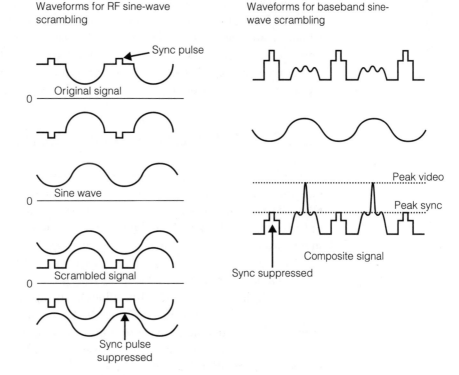

Waveforms for RF sine-wave scrambling

Sync pulse

Original signal

0

Sine wave

0

Scrambled signal

0

Sync pulse suppressed

Waveforms for baseband sine-wave scrambling

Peak video

Peak sync

Composite signal

Sync suppressed

tal sync for the channel to be scrambled are applied to the RF or IF signal suppressing the sync pulse, usually at 6 to 10 dB.

Some signal scramblers have the amount of suppression selectable at 0 (no suppression), 3, 6, or 10 dB. Suppressing the horizontal synchronizing signal fools the television receiver's sync separator circuits, thus preventing the set from locking the picture.

Other methods provide increased signal security because they further changed the video signal. A form of video inversion can be successfully used in conjunction with the gated sync method. Inverting the video signal (between horizontal sync pulses) interchanges the white and dark portion of the picture, resulting in an image like a film negative. Unscrambling is more complicated and hence more expensive to cable operators who think the higher level of security is worth it. Zenith Electronics Corporation used this type of signal scrambling on its Z-TAC system.

Because digital television is upon us, digital encoding with highly complex encrypting methods will offer cable operators extremely high levels of signal security. Two methods of digital encoding are line shuffling and line dicing. Of course, by mentioning "lines" we are referring to NTSC video lines of digitized video signals. Digital television is not simply digitized NTSC video signals; it is digital data generated by digital television cameras and equipment. Therefore, the security-coded algorithms are encoded at the point of transmission and decoded within the digital television receiver. Signal security is a high priority that protects the signals from unauthorized viewing.

6.25 The Interactive Subscriber Terminal

The subscriber terminal is connected to the drop cable at the simple entrance cabinet or to the structured wiring communications cabinet. Often if more than one set-top or computer modem device is to be installed at the entrance, drop cable will have to be divided by a power splitter. This means that enough signal level has to arrive at the initial input cable to be subdivided by the splitter. When structure wiring has been installed often there is a powered (amplifier) splitter that does not have any splitter loss. In general, the cable industry has adopted the open cable initiative as put forth by Cable Labs of Denver, Colorado. Cable Labs has been set up and supported by many cable operators and large multiple system operators (MSO). Cable Labs performs many research projects that improve cable service equipment and devices. As part of the Cable Labs open cable initiative, the specification for data over cable service initiative (DOCSIS) was developed. The set-top data gateway (DSG) protocol describes the signaling system between the set top and the headend. Downstream signaling is often in a space just above the FM band. Upstream signaling is in the 5–50 MHz sub-band and is often a digital PCM signal at 15 Kbps rate. Dialog as to the subscriber's request for service is placed on the upstream channel, and program contents on the downstream channel.

6.251   At one time it was proposed that the cable set top could be purchased through a local electronics outlet and put into service by the subscriber. This plan never went into operation, so at present the cable operating companies have such equipment available and provide installation and initiation of service. The installer of the cable service has to connect the cable plant at the subscriber tap to the grounded entrance block or housing with the coaxial drop cable.

The drop is either aerial or underground as determined by the cable plant. The installer has to connect the drop to the set tops through a power splitter if required and then initializes each set top, usually with the aid of a laptop computer. Once the installation is completed and initialized, the subscriber can select channels and order programs using the set-top remote control. Some set tops can have extensive memory for storing requested programs for later viewing.

6.252 The use of digital set tops with DSG protocol provides a certain amount of signal security. The use of a high level of signal encryption as opposed to simple scrambling and decoding makes theft of service extremely difficult. However, most cable operators don't underestimate the power of hackers. The flow of message traffic (signaling) between the headend and the set top monitors the system's program delivery quite well. Set-top devices need to be in constant dialog with the cable modem termination system (CMTS) at the headend. This CMTS also provides message dialog between the cable modem providing Internet service to the subscriber's computer. The computer communication data speed is several times the speed of the telephone method using the digital subscriber line (DSL) or the asymmetric digital subscriber line (ADSL) service and many times faster than the telephone system dial-up modem. High-speed Internet service is offered by the large MSOs using, for example, the Road Runner system. Fast data speeds allow the transfer of large data files consisting of pictures, games, and graphics. Such data service provides significant additional revenue to the cable operator.

6.523 Large system operators that have upgraded their cable plant expanding the upstream data capabilities are offering voice telephone service over the Internet (VOIP). Since telephone service is full duplex, the upstream or reverse plant has to be as good as the downstream plant. Cable operators that have expanded their upstream data capabilities are positioned well to offer VOIP and DOCSIS services.

The subscriber installation work has to be careful and adhere to high standards. The installing technician now has to install digital control set-top devices for television services and computer cable modems for the Internet connection. These technicians are going to have more training in order to make the service installations and do the necessary diagnostic work when difficulties occur.

Installing the modem usually consists of using a laptop computer or a CD-ROM to provide the initializing information to the subscriber's computer terminal. This program installs the necessary drivers used to communicate to the modem. Some subscribers are more computer literate than others and so understand what the installing technician is doing. Other subscribers often become bewildered and confused, so when troubles occur they call the cable operator. Most cable operators' customer service representatives can solve the problem over the telephone. This eliminates the on-site service call. Many system operators provide training for their technicians, A+ certification for example, which includes a working knowledge of PCs with emphasis on the computer input/output control functions. Servicing at the customer level is extremely important in keeping connects and disconnects to a minimum. Remember that there is competition with some telco operators, which are also offering video, voice, and data services.

Summary

The importance of the subscriber interconnection to the cable system is stressed in this chapter. Without a proper connection, communication and services between the subscriber and the system will be compromised. Cable operators are becoming very aware of the importance of the subscriber drop in providing quality of service (QOS) to the consumer. This, in part, has been learned from the telephone companies, which regarded QOS of extreme importance for many years. Only during extreme disastrous weather conditions or highway accidents did an interruption of telephone service occur. Most cable operators are aware of the importance service quality has in obtaining and keeping subscribers connected.

The service drop in comparison to the rest of the cable plant is very fragile. Metallic corrosion, ice, moisture, and the forces of wind over time affect the quality of the drop and hence subscriber service. Cable operators should examine the subscriber drop and measure and record signal levels every time a technician calls at a subscriber's home for whatever reason. Proper grounding and bonding procedures are very important to protect the drop from power and lightning problems.

The set-top terminal is the device that the subscriber uses to interact with the cable system. All requests for service and the service signals pass through this device. The complexity of this device varies from system to system. Even after a number of years, many systems are still using standard selectable set-top converters. Modern systems that have upgraded to high-speed digital bidirectional systems use sophisticated set-top devices that contain memory for storing ordered programs to be viewed at the subscriber's discretion. Also, the set-top device performs the decoding or descrambling function, which is an integral portion of signal security. Signal-security methods vary from the very fundamental process of signal trapping (denial) of nonsubscribed signals, such as premium programming. This method is expensive because nonpremium program subscribers had to have a pole-mounted trap installed. Signal-security methods improved to a scrambling/descrambling method usually employing an addressable converter. The scrambling process is performed on a channel-by-channel basis at the headend. The subscriber's converter is directed by the data received from the headend to unscramble the proper channel.

Many large system operators use even more sophisticated methods for encoding and decoding program channels. Because many such systems offer analog programming as well as digital programming, different methods for each are employed.

Recognize that the subscriber-to-system interface is extremely important. It is at this point the customer has face-to-face contact with a cable system installer or a maintenance technician. Telephone company installers were a well-mannered and professionally trained group who usually created a good impression of the company to the customer. They set a good example for the cable companies.

Questions

1. Identify the point for a cable drop where the signal level should be first measured.

2. Identify the point inside a subscriber's home where the signal level should be measured.

3. Where is the place in a subscriber's service drop that an electrical surge protector should be placed?

4. List in order the points in a subscriber's service drop that should be inspected for proper installation quality.

5. If the signal level is low at the ground block, describe the next step in the troubleshooting procedure.

6. If a proper signal level is observed at the ground block and it is low at the converter or set-top terminal, where would be the most likely place to check?

7. If an underground drop from a pole to a subscriber's house indicates it has been cut (no signal at ground block), what should be the repair procedure?

8. True or false: A low signal level at the subscriber's set-top terminal can cause poor signal decoding of coded premium channels.

Problems

1. A signal level of +10 dBmV is recorded at the ground block. If there is one signal divider between the ground block and the subscriber's set-top terminal, what is the expected signal level at the set-top terminal?

2. When using a signal-level meter to measure the drop level for a system that uses channel 2 through channel 70, which channel frequency should be measured and why?

3. If the signal level at the tap port is +10 dBmV and at the ground block it is found to be −10 dBmV, what is the most common cause?

Cable Plant Testing and Maintenance Procedures

Objectives

After learning the material covered in this chapter the student will be able to

■ Identify the basic installer test instruments and explain their operation.

■ Explain the theory of operation for the spectrum analyzer and its application to cable television testing.

■ Describe the process of testing cable and passive devices for signal attenuation and insertion losses.

■ Explain the initial turn-on and signal level balancing procedures for new cable systems.

■ Describe the use of testing a cable distribution system for signal leakage.

■ Describe the process of signal system sweeping and explain why this is a required test.

■ Explain the procedure for FCC proof of performance testing and describe the requirements for recording results for possible FCC inspection.

■ List and describe the proper tests for off-air, satellite, and microwave signal sources for quality.

■ List and explain why fiber-optical plant additions are used in state-of-the-art cable systems.

■ Describe testing procedures and equipment for test of fiber optical plant.

■ Explain the testing of cable system digital signals.

■ Describe cable system problems and the troubleshooting procedures used to identify and cure them.

7.1 Instruments and Measurements

The maintenance of a cable system is extremely important and necessary to provide quality service to subscribers. Adding two-way services for data communications requires constant monitoring to prevent noise and signal ingress. As long as the FCC holds cable systems responsible for RF signal leakage, cable operators will have to work hard to maintain a tight cable plant. This is also important in reducing reflections and keeping the signals relatively noise-free. In order to perform the required maintenance procedures, calibrated test equipment in sufficient quantity should be available to the technical staff.

7.11 Signal-Level Meters

The most common piece of test equipment used by cable technicians is the signal-level meter (SLM), which is used to read the magnitude of an RF carrier. Early signal-level meters were essentially tunable voltmeters with electromechanical meters indicating the level in microvolts. Because converting the voltage value to dBmV was more appropriate for cable technicians, the meters were calibrated to indicate the level in dBmV. The peak detector is used by television SLMs and is appropriate for measuring peak carrier amplitude that occurs at the horizontal sync pulse tips. A basic dual conversion SLM is shown in Figure 7-1. As shown in the figure, the circuit is basically a dual conversion receiver with a peak detector driving either an analog meter movement or digital LCD display screen. Early meters used the standard analog meter movement. The bandwidth of such a meter was quite narrow and on the order of 400 KHz.

7.111 Throughout the years, the SLM has gone through many improvements in level accuracy, bandwidth, tuning stability, and display technology. From an operational standpoint, because these instruments are

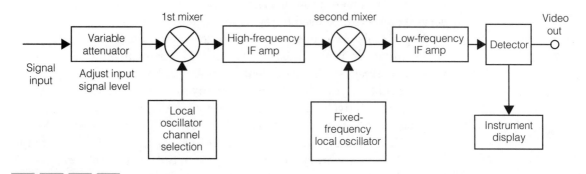

Figure 7-1
Basic dual-conversion signal-level meter

portable, battery life has been extended as well as a reduction in weight. Years ago, climbing a pole and carrying an SLM was hard work. Present-day meters produced by such manufacturers as Sencore, Wavetek, and Sadelco, to name a few, are a bit larger than a hand-held digital multimeter. Accuracy and usability of current SLMs are excellent.

Early meters employed a tuning knob and dial to select the channel to be measured. Also, a switched attenuator was used to adjust the signal level near the center of the meter movement. This attenuator was usually a step attenuator in a number of dB-per-switch positions. Also, some instruments needed to be adjusted for frequency response variations for selected frequency bands. Present-day meters using chip technology have extremely flat responses, and no band calibrating is necessary. Most meters used today have batteries that can easily be recharged overnight.

7.112 Most cable systems operating today equip their installation crews with an SLM to check the tap level at points across the band and at the subscriber's terminal inside the home. This information should be added to the subscriber's file every time it is made. The first time, of course, is at initial installation and should be measured and recorded at each trouble call. This information can be valuable in tracking system problem areas.

Another piece of test equipment the installer should have is a good, rugged digital multimeter (DMM), which can be used to measure the cable system power and the leakage voltage from the ground or between the subscriber's equipment. Cable systems offering telephone service will have to carry the necessary power for the telephone set on the cable system. Such voltage can be measured by the DMM.

The installation crew is required to carry a leakage detection instrument. Measurement of leakage is a part of the ongoing signal-monitoring program. Documentation of the tests performed by the installation and service crews can be used to report leakage information.

Proper tools are a must for installation crews. Such tools include the cable connector preparation tools, such as stripping and crimp tools, which are necessary to install leakage and weatherproof connectors. Correct installation procedures can save many future service calls.

Another relatively inexpensive piece of equipment that can provide valuable information is a portable GPS receiver. Subscriber addresses will have a GPS location entered in the plant database. This receiver should have sufficient accuracy (operate in differential modes) in order to differentiate between close subscriber addresses. Software of a table lookup type can be useful in doing leakage location studies as well as installer/service crew locations using the GPS information.

7.113　Many present-day SLMs can perform several valuable functions. Such functions include determining the signal level of television video/audio carriers, signal-to-noise (S/N) ratios, and hum and low-frequency disturbance measurements.

Meters made by several manufacturers use a screen type LCD display. Displays of this type are driven by digital circuitry, making it simple to store screen displays in digital memory. Periodically, the memory can be loaded into a PC for analysis and storage. Automated test equipment operating in this manner enables the technical department of a cable company to keep accurate records of plant measurements. Plant test records stored on disks or CD-ROMs can be retrieved quickly for analysis and also have the added benefit of not taking up much space. Many SLMs with screen-type LCD displays approach the capability of a spectrum analyzer and are often referred to as the poor man's spectrum analyzer. An example of a screen layout is shown in Figure 7-2.

7.12 Spectrum Analyzers

For RF signal measurements, there is nothing quite like a modern spectrum analyzer. Like the SLM, spectrum analyzers have gone through several major improvements. They improved in accuracy and in number of features, and several more manufacturers are offering them. Both Tektronix and Sunrise Telecom offer spectrum analyzers to the cable indus-

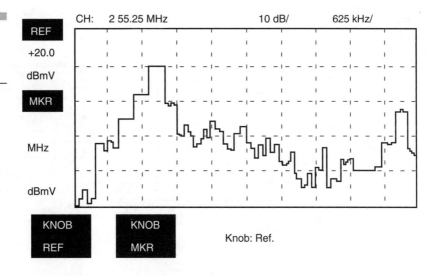

Figure 7-2
LCD screen of TV channel 2 digital display

try with sufficient test modes and accuracy to perform all the necessary tests required by the FCC for system proof-of-performance tests. Some of the features cable technical personnel consider as important as accuracy and performance are simply whether an instrument is lightweight and battery powered. This makes using the equipment in the field a lot more practical. Digital data storage as well as digital data setup of the instrument is also important to large cable systems that require a lot of measurements. A test setup using such features is shown in Figure 7-3.

7.121 The spectrum analyzer contains some of the circuitry contained in the SLM. The display used by a spectrum analyzer is a cathode ray tube or a bright LCD screen. The screen is driven digitally. Spectrum analyzers offer an analog display mode as well as a digital one, giving the operator the best of both worlds.

A block diagram of an elementary spectrum analyzer is shown in Figure 7-4. This block diagram indicates that the width of the measured signal is essentially the width of the resolution bandwidth filter, which is the last IF amplifier in the spectrum analyzer signal chain. The instrument is also swept through the frequency band at the same rate of the display, following the horizontal power contained in the resolution bandwidth circuit.

A screen display of a television signal channel is shown in Figure 7-5. In this case, the instrument controls are set to display the entire signal across the 6 MHz band. As mentioned earlier, spectrum analyzers on the

Figure 7-3
Automated
measurements and
data logging

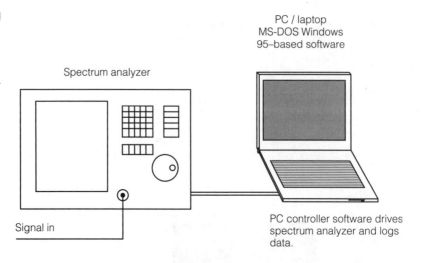

Spectrum analyzer

PC / laptop
MS-DOS Windows
95–based software

Signal in

PC controller software drives
spectrum analyzer and logs
data.

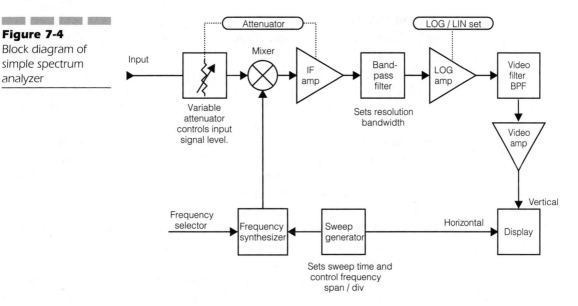

Figure 7-4
Block diagram of
simple spectrum
analyzer

Attenuator

LOG / LIN set

Input

Mixer

IF
amp

Band-
pass
filter

LOG
amp

Video
filter
BPF

Variable
attenuator
controls input
signal level.

Sets resolution
bandwidth

Video
amp

Vertical

Frequency
selector

Frequency
synthesizer

Sweep
generator

Horizontal

Display

Sets sweep time and
control frequency
span / div

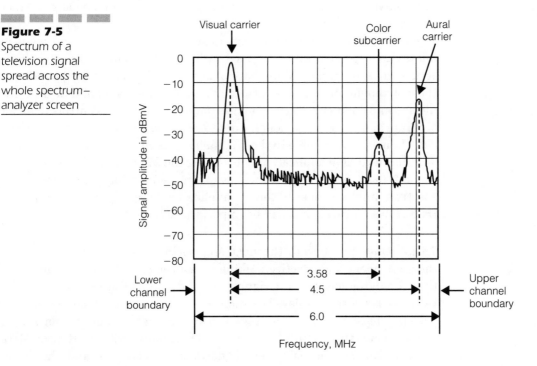

Figure 7-5
Spectrum of a
television signal
spread across the
whole spectrum—
analyzer screen

Visual carrier

Color
subcarrier

Aural
carrier

Signal amplitude in dBmV

0
−10
−20
−30
−40
−50
−60
−70
−80

Lower
channel
boundary

Upper
channel
boundary

3.58
4.5
6.0

Frequency, MHz

market today offer many options, and many manufacturers offer analyzers geared to cable television applications.

7.122 Spectrum analyzers with special cable television features are available from several manufacturers such as Sunrise Telecom, Tektronix, and IFR. A technician or engineer can program these instruments for various cable applications. Digital signal output jacks on the rear panel can be connected to a PC or a printer. Spectrum analyzer measurements can be printed out and placed in a notebook or can be saved and transferred to a PC later. When many measurements are needed, such automation features can become a time-saver.

A TEK-2714/15 spectrum analyzer is a made-for-TV application. This instrument can display in either analog or in digital mode. It can have its controls set by a PC and also save digital data measurements. The use of a laptop PC connected to this instrument can save a lot of time when many measurements have to be made. Instruments with such time-saving features for instrument setup and data, of course, cost more and are well worth the expense, considering the time saved. Systems with data storage and analysis software can save further time and effort in keeping track of system performance.

The controls of many analyzers today use push buttons and what is known as soft-touch controls to set up the instrument to make the prescribed measurements. Many have a data entry keypad to further control the instrument. As might be suspected, it takes quite a bit of time for technicians and engineers to learn all the major features and controls to make accurate signal measurements. *Cable Television Proof-of-Performance* by Jeff Thomas is applicable to the HP 8590 series spectrum analyzers. Also, TEK has available application notes using the 2714 spectrum analyzer. It would be a good idea for technical people contemplating the purchase of a spectrum analyzer to send for the application notes and make a study of the various features that will be important to their needs.

7.123 Once purchased, the next step is to learn to properly operate the instrument and practice making measurements. Basically, any spectrum analyzer measures a signal's amplitude in dBm or dBmV on the vertical axis of the display. The vertical axis is usually scaled in 10 dB or 2 dB per division. Some analyzers also have a linear scale. The horizontal axis indicates the signal's frequency and is usually scaled in KHz or MHz per division. The measurement bandwidth is selectable and should be properly chosen based on the speed of the display sweep. If the resolution bandwidth filter does not have time to respond to the signal due to the sweep speed, an error in signal level will result. Therefore, in the past, some manufacturers of spectrum analyzers locked the sweep time controls to the bandwidth control so the problem could be avoided. For those needing to have control, the controls could be unlocked in order to make the needed measurements. Present-day spectrum analyzers are microprocessor-controlled to avoid most of these problems.

As a rule of thumb, you should vary the resolution bandwidth if the signal amplitude increases the maximum amplitude settings. This topic is also explained in Jeff Thomas's *Cable Television Proof-of-Performance*. The information on this subject is also covered in several application notes by TEK. Many cable technicians using the TEK-7L12/7L13 spectrum analyzers are familiar with this characteristic.

Essentially, the spectrum analyzer measures signal amplitude and frequency to a high degree of accuracy. Also, by observing the spaces between signals, noise and spurious signals can be measured. When a spectrum analyzer is placed in the zero-span mode, it acts simply as a tuned receiver, and the modulation components can be observed. Essentially, in the zero-span mode, the analyzer acts like a SLM with an oscilloscope readout. Now the detected signal can be displayed in the time domain.

For television channels, the signal displayed will be the base-band video signal. The IF, or the resolution bandwidths, has to be wide enough to pass the video signal. By displaying the video signal with a slow sweep speed, hum and low-frequency disturbances can be observed. For the hum measurement, the analyzer has to be placed in the linear vertical scale mode and the operating instructions pertinent to the particular analyzer should be followed. A high-quality spectrum analyzer can be used to make all the measurements required by the FCC proof-of-performance RF test. Specifications for a spectrum analyzer used for cable television systems are given in Table 7-1.

Other features that are helpful in using the analyzer to make accurate measurements are

- Frequency and amplitude markers
- Analog CRT display or LCD display
- Choice of sweep speeds in zero-span mode (10 for full span)
- TV sync trigger
- Fast Fourier transform (FFT) circuit on detected output
- FM demodulator
- Television picture and sound on CRT or LCD display
- Negative peak detection

Chapter 7

Table 7-1

Minimum
Specifications for
a Cable System
Spectrum Analyzer

Frequency range:	10–1000 MHz, usually 9 kHz–1800 MHz
Frequency spans:	0, 100 kHz–1000 MHz, usually 100 MHz–1 kHz 1-2-5 steps
Frequency accuracy:	± 200 Hz, typical counter accuracy 5×10^{-7}
Relative amplitude accuracy:	± 2 dB over frequency range, flatness ± 2 dB
Maximum input level:	1 watt (damage level) typical, $+20$ dBm max or 68.8 dBmV
Sensitivity:	minimum signal level -60 dBmV minimum CATV signal
Noise floor:	-60 dBmV with narrowest resolution BW
Internal distortion products:	$\leqq 60$ dBc with analyzer's mixer input 10 dBmV. Various analyzers may state this differently
Resolution bandwidths:	1 kHz, 3 kHz ,4 MHz video Bw; usually analyzers have res Bw of 5 MHz, 1 MHz, 300 kHz, 100 kHz, 30 kHz, 10 kHz, 3 kHz, and 1 kHz
Video bandwidths:	equal to resolution bandwidths
Input attenuator:	variable 0–60 dB (10 dB steps), some analyzers have multiple steps of 2 dB each
Input preamplifier:	(internal or external) gain >20 dB and noise figure < 7 dB
Input impedance:	75 ohms

7.13 Cable and Passive Device Testing

The signal transport method used in the cable television industry is the coaxial cable itself. This cable is connected together using active devices (amplifiers) and passive devices (taps, couplers, splitters, power inserters). Testing of these devices should be done before installation to ensure that no faulty components get installed in the system. A program of testing the devices before installation is termed admittance testing. The test procedures for such equipment are for frequency response and loss.

7.131 Loss at the upper and lower frequency bounds for cable and passives are basic and most important. Loss measurements are simple to make and are shown in Figure 7-6. A fixed level of signal is injected and simply measured at the output and should be performed at the lowest and highest frequency. The cable or device can pass the signal at these two frequencies, the upper and lower frequency bound. Many systems make only this measurement for the admittance test. Comparing the loss figures with those given by the manufacturer can give an indication of the length of cable on the reel.

7.132 A signal sweep test can provide even more information about cable and passive performance characteristics. The results of a sweep test tell the loss of the device over the operating frequency band. Any differences in loss as a function of frequency can be observed. This measured information can be compared against the manufacturer's performance specification that verifies equipment as satisfactory. Figure 7-7 shows a sweep setup for cable testing.

Figure 7-6

Output port A is tested at f_{LO} and f_{HIGH} and levels measured by the signal-level meter

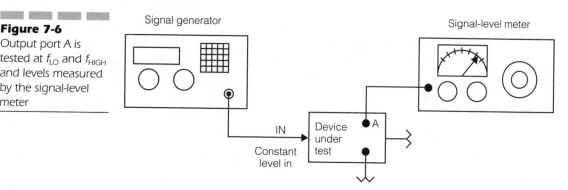

Figure 7-7

Sweep testing of coaxial cable

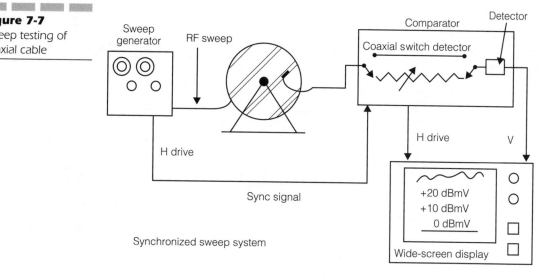

7.133 Another test for passive equipment is a return loss test. Return loss in dB is simply equal to 20 times the logarithm of 1/reflection coefficient. This test is performed using a return-loss bridge driven by a white-noise signal source and is shown in Figure 7-8. This test basically is a measure of the characteristic impedance variations from the matched condition. Variations in impedance match can cause signal amplitude variations as well as television signal ghosting in extreme cases.

7.134 Another easy and interesting test to make for cable is using a time domain reflectometer (TDR) to measure the length of the cable on a reel. There are several TDR instruments on the market today that operate in the digital domain, having an LCD screen that indicates the distance to a fault. This distance will equal the length of the cable on a reel if the end is open or short-circuited. For testing the length on a reel, it is usual to leave the end open (a matter of convenience) and measure the distance to the fault (a cut cable). Any abnormalities observed along the cable length will be shown on the LCD display and can be measured as dent or crush. This TDR measurement is often used alone as an admittance test for cable or used in conjunction with the simple loss test. The test setup and an example of some screen displays is shown in Figures 7-9a and 7-9b. A TDR displaying the waveform tells the most about cable problems. Some lower-priced TDRs with digital readouts indicate the distance to the first

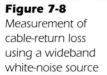

Figure 7-8
Measurement of cable-return loss using a wideband white-noise source

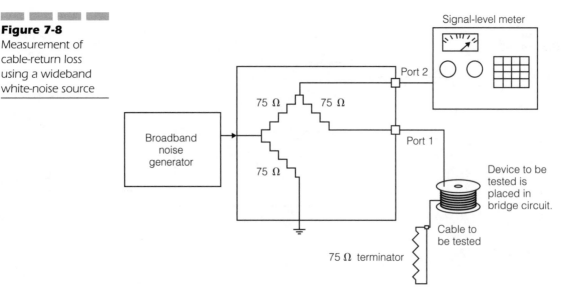

Figure 7-9a
Test for length of
cable on a reel

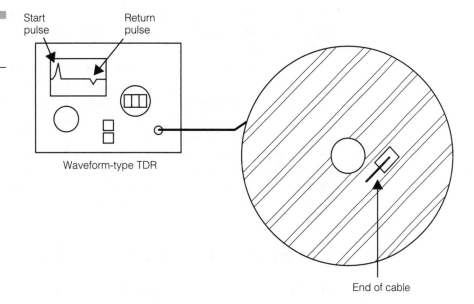

Figure 7-9b
Measurement of
distance to cable
fault for buried plant

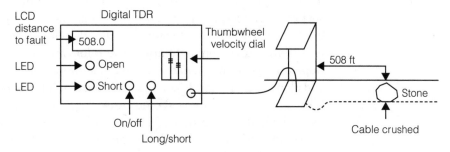

fault. An LED indicator points out the nature of the fault, such as an open
(cut) cable or a short-circuited cable (crushed).

7.2 Cable System Tests and Measurements

Cable television systems are tested as mandated by the FCC, which is
referred to as the proof-of-performance test. Cable systems are continu-
ally tested for cable system signal leakage. Cable systems that are completely coaxial and are driven by amplifier cascades have a much more

extensive problem keeping cable signal leakage under control. Also, the long cascades of trunk amplifiers should be checked periodically for noise and distortion problems.

Cable systems that use fiber-optic technology where the trunk cable is replaced by an optical fiber carrying both the downstream television signals as well as upstream digital data signals have a lot less signal leakage problems, because leakage can exist only in the coaxial feeder plant. Also, the coaxial feeder plant usually consists of short RF amplifier cascades. Many systems limit the amplifier cascade to five. These systems also enjoy the fact that cascade noise and distortions are minimal or unobservable. Still, as long as there is coaxial cable plant, RF leakage has to be controlled and undergo the FCC proof-of-performance testing.

7.21 System Turn-On and Balancing

The first test any cable system should have is the construction contract compliance test, which is essentially the same as the FCC proof-of-performance test. Passing this test means that the system fulfills the contract specification that it complies with FCC standards.

Initially, at system activation, the pilot frequencies used for amplifier slope and gain are turned on. Next, the amplifiers are adjusted for gain and slope. Other signals can also be injected, but the pilots are necessary. Now the activated plant should be leakage-tested using the pilot carriers. A leakage test transmitter can be placed at the headend, which has an easily recognized signal that can be detected by portable leakage test receivers. Several manufacturers have such devices available to the industry. Use of these special pieces of equipment can speed up these tests, and the cable operator can use this equipment as part of the ongoing leakage control program. Leakage testing done at this time assures that the new system is signal-tight and ready for initial turn-on and balance.

7.211 The initial balance procedure begins at the headend and progresses down each major trunk leg, balancing each trunk amplifier for slope and gain. At each bridging point, the feeder legs are then set and adjusted so when the end of the trunk leg is reached, all the connected feeders are active. Many systems will connect subscribers as each feeder leg is activated. This helps company image as well as cash flow.

7.212 Once the signal levels are set, the system can be frequency-swept. Companies such as Sunrise Telecom and Acterna have specialized system-sweep testing equipment that provide accurate results and are relatively easy to use. Many companies depend on the construction contractor to sweep the system, with the work monitored by one of the system's technicians. Some cable operators do not purchase such sweep equipment but may rent or lease it when needed. A system sweep setup is shown in Figure 7-10. Large companies with several systems often have a sweep system that is used on a rotational basis.

7.22 Proof-of-Performance Testing

The FCC proof-of-performance test is mandatory, and all systems with coaxial plants have to comply. The commission adopted new regulations in 1992 calling for higher levels of performance. The tests are the same, but the performance standards are raised. Results of the proof tests have to be carefully documented and placed in the cable operator's technical file. Essentially, the FCC reinstated and revised the Part 76 technical standards pertaining to biannual measurements at widely separated test points in a cable system's distribution plant. One measurement point should be at the longest amplifier cascade. For the first three years starting in February 1992, only RF measurements are required. From then on, demodulated base-band video tests have to be performed on a three-year cycle. The RF measurements consist of signal measurements performed at the test points as well as a systemwide cumulative leakage index of 64.

Figure 7-10
Noninterfering
system sweeping

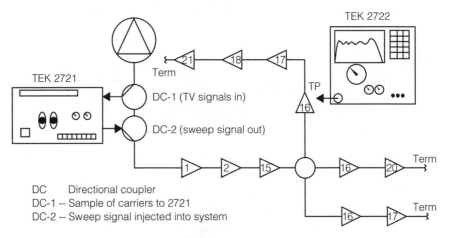

Automatic sweep system using TEK 3721/2722

7.221 The FCC regards the system leakage tests as extremely important. In the past, it was feared that cable signal leakage occurring in the air-safety and navigation frequency bands could constitute a hazard to air safety. The Federal Aviation Administration (FAA) pressured the FCC, which in turn pressured cable operators to clean up their plant. Although it is extremely doubtful that any serious conditions are caused by cable system leakage, some cable operators reported fewer plant problems and service calls when the leakage problems were fixed.

Many cable operators have taken advantage of equipment made specifically for finding cable system leaks. Essentially, a transmitter with an easily recognizable audio tone is placed in the headend and combined with the television signals. Receivers in the service trucks can monitor this frequency, which aids in finding the source. These receivers are sensitive and can pick up leakage from the pole plant. Once the area of the leak is found, technical personnel with hand-held receivers pinpoint the problem. An ongoing leakage control program integrated with the maintenance program means that passing the cumulative leakage test for proof of performance is likely to happen. The accepted method for CLI testing, however, is done with a dipole antenna and a signal-level meter. This procedure is discussed in Chapter 2. The leakage specifications are given in Table 7-2.

Most cable systems use the on-the-ground testing methods for monitoring and repairing system leaks. However, some large systems covering many areas may elect to do an aerial test. The aerial test is generally administered by specific contractors, who have aircraft outfitted with monitoring equipment and special antennas. This method is known as a flyover method. Some cable operators feel that this is a more realistic

Table 7-2

FCC Signal Leakage Limits

Frequency Band (in MHz)	Field Strength (in MV/meter)	Test Antenna/ Distance from Cable Leak (in feet)
<54	15	100
54–216*	20	10
>216	15	100

*The range 108–137 is the aeronautical frequency band. It is a good idea to test a carrier in this band for leakage.

measurement, because the crux of the leakage problem is concern for the interference of air-safety communications. Equipment in the aircraft can provide data through the measurements to map hot spots of radiated signals. Now the system technical personnel can converge on these areas and repair the actual leaks. Documentation of measurement data, time and date of measurement, calibration of instruments, and calculation of CLI is important information to be filed in the system's technical file.

7.222 The signal test requirements for RF and base-band video signals are stated in detail in Part 76 of the FCC regulations. The RF signal tests are conducted twice a year. Therefore, multiple system operators (MSO) usually have a proof-of-performance crew going from system to system on a rotating basis. Technicians doing this work get good at doing it through the constant practice.

The FCC rules state that systems with fewer than 1,000 subscribers are exempt from the proof-of-performance test. Systems with 1,000 to 12,500 subscribers have to perform the test at six widely separated test points in the system, and two of which have to be at the points most distant from the system input (headend or signal source node). One test point is added for each 12,500 subscribers. Results of the proof-of-performance test must be kept in the company files for a period of five years. Records can be either in notebook or file-folder form or stored on floppy disk. If a representative of the FCC calls on a company, the five-year records can then be printed out and presented for study. These records have to include the names of the personnel who made the tests and their qualifications, as well as equipment calibration data. Many systems have equipment such as spectrum analyzers, frequency counters, and signal-level meters, which must have a calibration check every two years.

Essentially, the FCC regulations require the following basic categories of the testing:

- Documenting the names and qualifications of personnel performing the tests.
- Documenting test equipment used, with the last calibration test and date performed.
- Testing the signal level at subscriber's terminal to make sure the signal is adequate to deliver the service.
- Testing the quality of the signal for frequency, noise, and distortion products. Such testing indicates the system is producing a high-quality signal to the subscriber.

■ Providing proof of compliance that the system is controlling leakage below any level that could interfere with over-the-air licensed services.

The personnel performing tests have to demonstrate competence through school training or correspondence school training certified through a certificate or diploma. An SCTE training program certifies a technician's or an engineer's competency to perform the FCC tests. Some cable companies have on-the-job training programs that could be acceptable to the FCC.

The equipment used should be periodically calibrated by the equipment manufacturer's service department or by an independent test and calibration laboratory. The date of the calibration service has to appear in the FCC proof documents.

The testing of signal level at the subscriber's terminal refers to the input to the subscriber's TV set or the set-top converter output. Some operators perform a test on the input/output characteristics of an arbitrarily selected converter and use this information for this test. Now the signal level of the tap port can be used for 100 ft of drop cable (3 dBmV). Some cable technicians will take a 100-ft piece of cable and connect to a test point tap port. Connecting this cable to a converter in the test vehicle allows this measurement to be accurately made.

The present FCC proof-of-performance tests in force today are for NTSC television signals. When digital transmission arrives, the FCC no doubt will produce new signal tests and specifications for compliance. For the present, we will all comply with the present FCC proof-of-performance tests.

It is assumed by now that all cable systems have the latest rules in their possession, and this book discusses helpful tips and techniques. The signal level at the subscriber's terminal can be measured using either a signal-level meter or a spectrum analyzer. Some systems connect to the set-top converter and then observe the level at the output of the converter, noting any signal-level variations as the converter is tuned through the cable channels. This, most likely, is not acceptable by the FCC, but signals that are low in level can be quickly spotted and measured.

Other signal-level tests are to be made over a 24-hour period, which can indicate any signal-level variation between daytime (warm) and nighttime (cool). The FCC specification states the video signal level will not vary more than 8 dB within any six-month interval, which must have four tests in six-hour increments (over 24 hours) during July or August. For January or February, the video carrier level will not vary more than

3 dB within a 6 MHz carrier separation, or 10 dB on any other channel. This measurement is performed on the visual carrier of each cable channel. Many systems use three technicians working eight-hour shifts to accomplish this test, while another test crew will visit the six or more test points making the signal-quality tests. Proper documentation of tests can simply be written on preprinted forms or stored on a laptop computer floppy disk. Some computer-oriented systems use a software program to compile and/or print the data.

Signal-quality tests concern the signal spacing of the video and audio carrier. To do this, usually the visual carrier frequency is measured and then the audio carrier; the difference should be 4.5 MHz $\pm$ 5 KHz. This measurement can be made using a TEK-2714/15 or HP 8590 spectrum analyzer.

Before these newer spectrum analyzers were available, a tunable down converter was used with a digital frequency output indicating the visual/audio frequency offset (4.5 MHz). Another manufacturer had a tunable signal stripping analyzer that could read the visual carrier frequency and the audio carrier offset. Such equipment provided the proper accuracy but was more difficult to use and a bit more time-consuming.

The television signals must be tested for noise and low-frequency disturbances (hum). The carrier-to-noise ratio is a particularly important specification. Usually, this appears worse at the longer cascades because this noise builds up with the number of amplifiers in cascade. This subject is covered in detail in Appendix E. Since the specification is stated as greater than 36 dB, most cable operators try to keep the carrier-to-noise (C/N) ratio at the subscriber's terminal at the longest amplifier cascade to 46 to 47 dB. Subscribers with large-screen TV sets find the 36-dB specification allows for the noise to be objectionable. The cable industry should strive to increase the specification to 48 to 49 dB C/N. With this specification, systems can be more competitive with the direct digital broadcast services (DBS) satellite-delivered services. This test can be made with some of the present-day signal-level meters most cable operators use.

Also, such level meters can make the hum and low-frequency disturbance tests. This measurement can also be performed using a high-quality spectrum analyzer. When using any spectrum analyzer, a pre-selector filter is needed to protect the input of the analyzer from other signals that can be large enough to overdrive the analyzer's mixer. Strong signals entering the mixer can cause the analyzer to produce spurious signals, rendering the measurement inaccurate. Pre-selector filters are available from Acterna and Trilithic, to name a couple. Proof-of-performance test gear usually has pre-selector (TV channel band pass) filters in a cabinet, with each one tuned to the test channels. Such an arrangement of equipment is shown in Figure 7-11.

System flatness or frequency response tests can demonstrate that the distribution system is providing the proper signal level throughout the spectrum or downstream frequency band. A sweep test using a simultaneous sweep transmitter at the headend and a sweep receiver at the test point is the usual test that provides this information. Systems that are fully loaded can be tested using a spectrum analyzer or one of the new signal-level meters that display visual and audio carrier levels across the band. Also, the audio carrier level should be at least 13 dB below the visual carrier level, the usual being 15 dB. This prevents sound beats in the picture of the upper video carrier.

In-band frequency response addresses the frequency signal flatness within the 6 MHz channel. The regulation states that the response should not vary ±2 dB from 0.75 MHz to 5.0 MHz above the lower channel frequency boundary. This test is actually a sweep response test for each channel. Usually, this pertains to modulators and not signal processors. More elegant cable systems that use demod/remod schemes and have the channels phase-locked to a comb generator, providing the incrementally regulated carrier (IRC) or harmonically regulated carrier (HRC), should test the modulators carrying the off-air broadcast television services. A video sweep generator will provide the signal to the modulator and the receiver will detect the RF output response over the 6-MHz band. An in-band sweep test is shown in Figure 7-12, including an example of the measurement. This type of test is conducted on the equipment at the headend. The spectrum analyzer can act as the sweep receiver for the RF output of the modulator. If an RF sweep generator provides a narrow RF

Figure 7-11

Distortion test of the cable system

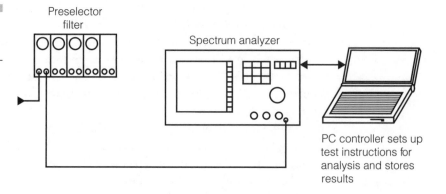

Preselector filter

Spectrum analyzer

PC controller sets up test instructions for analysis and stores results

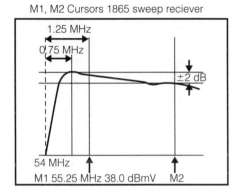

Figure 7-12

In-band sweep-
response test

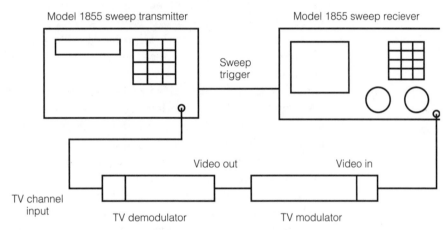

signal 6 MHz wide, then a processor can be sweep-tested over the channel bandwidth. Acterna manufactures the 1855/65 sweep system, which can be used to do all the required system sweep tests. This equipment operates on a battery with an internal charger that provides several hours of operating time. If the service vehicle has a commercial power source in the form of either a rotary converter or solid-state converter, the sweep receiver can be charging while driving to the next test point.

As stated earlier, the state-of-the-art spectrum analyzers with an accurate frequency counter built in can be used to test the cable system aural (sound) carrier frequencies to the FCC required accuracy of 4.5 MHz ± 5 KHz. This is measured at the subscriber terminal. Usually, this is measured at the headend and should not be affected by the distribution system or the set-top converter. Other specified frequencies that have to be tested are the channels that operate in the aeronautical frequency bands, which are 108–137 MHz and 225–400 MHz. Specifically, cable channels operating in these bands have to be offset from certain aeronautical carrier frequencies by certain offsets. An offset of 12.5 KHz ± 5 KHz offset or 25 KHz offset + 5 KHz is required of cable channels operating at power levels of 10^{-4} watts, which corresponds to −10 dBm or about +39 dBmV. This level is within the level of line extenders or distribution amplifier's output. Therefore, cable systems have to comply with this offset regulation.

The proof-of-performance test logs should be preprinted with the channel frequencies to be tested also listed, including the offsets so measured data can be entered in the next column. This test is done at the headend and is performed first before other tests can proceed. Frequency testing can be done with a signal-stripping tunable frequency counter or preferably one of the new state-of-the-art spectrum analyzers containing an accurate frequency counter. Frequency accuracy for such instruments should be at least four parts in 10^{-7}. As any frequency counter gets older, the time base system deteriorates. This is referred to as aging and is usually given in the instrument specifications. For example, if the instrument aging specification is 1×10^{-7}/year, then if the instrument is going to be used to measure 1 GHz (upper frequency limit), then the measured frequency can drift ±100 Hz. This result is calculated by multiplying the 1 GHz by the aging specification. Regardless of the instrument used to make the frequency tests, the operator should study the instrument's operating manual and practice making the measurements before doing the actual proof-of-performance tests.

In the past, the spectrum analyzer has been the instrument of choice for measuring signal distortions. Signal distortion breaks down into second- and third-order distortions. Composite second-order distortion (CSO) is made from the signals being carried on a cable system. All distortion levels have to be 51 dB below the visual carrier level when the system is not phase-locked (IRC, HRC) and 47 dB for phase-locked systems. These measurements should be averaged over time.

Digital storage in the spectrum analyzer can be very helpful. Most of us are very much aware of the calculations of system beat buildup due to the small amount of amplifier nonlinearity. Also, the buildup is a function of the number of channels carried on the system and the number of amplifiers in cascade. Present-day systems using fiber-optical trunking to distribution nodes typically carry many channels (approximately 100–110 channels) through short cascades of usually no more than five amplifiers and hence such distortions are well within specification. Because the cascade factor, as discussed in Appendix E, is 20 log n, n is the number of amplifiers in cascade, which calculates to five amplifiers to 14 dB. For amplifiers with a composite triple beat (CTB) of 68 dB, then for the five amplifiers cascade relates to $68 - 14$ or 54 dB which is 3 dB better then the allowed 51 dB specification. Therefore, the conclusion can be made that for systems rich in optical fiber, where the fiber-optical plant has replaced the coaxial cable trunk, the cascade number is reduced to five amplifiers with high quality amplifiers. The CSO and composite triple beat (CTB) specs are at least 70 dB. The likelihood of seeing any measurable distortion is quite remote. Also, such systems have reduced the number of miles of coaxial plant, thus reducing the possibility of cable leakage. However, the coaxial distribution part of the cable system operating at high-signal level will have to be diligently monitored for ingress and leakage.

7.23 System Maintenance Measurements

A system maintenance program is extremely important in providing good service to subscribers. More important is that many cable operators are adding, if they have not already, digital services to residential and business subscribers. In order to do so, the cable operator has to have an upstream or return path that is operating noise-free and ingress-free. The procedures vary from system to system, depending on the network topology. Some systems do a market survey first and then activate the reverse plant accordingly.

For all coax systems, activating the reverse system can be a difficult job. This is done by reworking sections of the plant, progressing back toward the headend. For sub-split reverse systems, the subscriber drops should have band-reject filters installed to limit signal input to the reverse system. Each section should then be forward-leak tested to locate any loose connectors where noise and signal ingress can occur. For systems with an optical-fiber trunk, the upstream reverse is optically transmitted from the fiber coax node to the headend. The reverse signals are the "T" sub-split reverse carriers with a digital modulation QAM-16, for example.

Many of us may wonder whatever happened to the mid- and high-split systems that were discussed a few years ago. If any were actually built, most likely their trunk systems have also been replaced with optical fiber. Such systems had much more bandwidth for upstream applications, but it should be obvious that a well-maintained system would give us fewer problems when activating the reverse system.

7.231 One of the simplest and most valuable tests is a system end-level test. To use such a test program, ground-level test points should be installed by running a drop from a tap connected near the last line extender (LE) amplifier at one of the system ends. If the cable system has a skeleton plant diagram, the end points where tests should be performed will be obvious. A sample of such a diagram is shown in Figure 7-13.

Systems using long cascades of trunk amplifiers can exhibit poor system flatness responses due to the amplifiers' characteristic signature. A slight dip, for example, in the mid-band can develop to a deep valley down the cascade. In the past, the amplifier manufacturers did not manufacture the amplifiers to tight tolerances, so the dips and small peaks varied differently, causing the effects to cancel. End-level tests on the system's carriers spaced throughout the spectrum will indicate any flatness problems before they become critical. Some systems elect to take end-level tests monthly and some bimonthly. Tracking the measured values through the seasons provides information on how well the thermal compensation and automatic gain circuits (AGC) are operating. These measurements are easily made with most present state-of-the-art signal-level meters with a keypad channel entry. Some enable the instrument to step through the programmed channels to be tested by the touch of a button. Analyzing the data can be made easier by having the data recorded on a disk and having analysis software provide graphical plots on histogram charts of signal-level variations.

7.232 Leakage testing is the other required maintenance program. A good leakage program aids a system in keeping the plant tight, keeping leakage, ingress, and noise to a minimum. Often a system will find connector problems before upstream data service problems develop.

Although the FCC requires leakage tests to be made using a test dipole, leakage test equipment sets made for finding system leaks are more accurate, easier, and faster to use. A calibrated leak can be made near the maintenance department and activated when the install and service trucks are leaving in the morning. White lines indicating the 10-ft distance and the drive-through lane can be painted on the parking lot. The trucks passing through can activate their equipment and check instrument calibration on their way out to work. After leaving, the equipment generating the calibration leak signal can be turned off. Continual testing for leaks

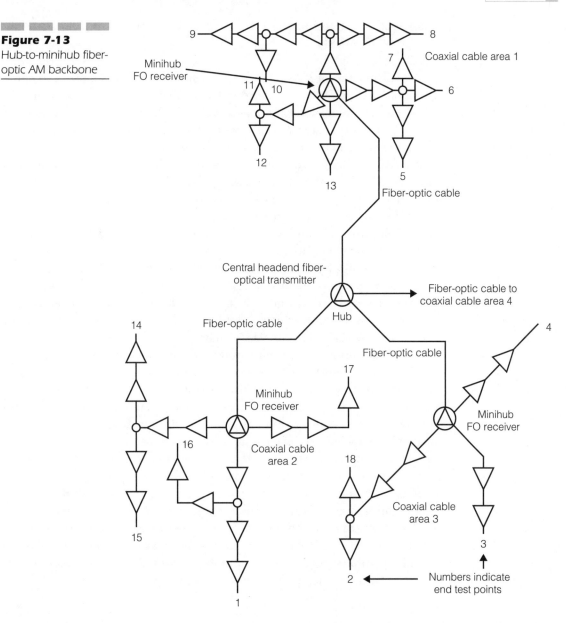

Figure 7-13

Hub-to-minihub fiber-optic AM backbone

and making needed repairs can ensure passing the CLI tests as required by the proof test. Some systems that have a plant skeletal map enlarge this map and place it on a wall in the maintenance office or garage. When leaks are found by the various install and line technicians, colored pins can be placed on this map. Then the work effort can be directed at the "hot" spots. Such techniques can increase efficiency and improve the leakage problem quickly.

7.233 Proper record keeping of test results can be a benefit in keeping the cable system in top-notch order. Logs of headend signal levels on each of the carried television channels can often indicate when a modulator or signal processor is starting to have problems. The headend signal levels are extremely important for coaxial cable trunk runs. At the beginning of a cascade, distortion and noise increases by the number of amplifiers in cascade. Headend signal level stability should be closely monitored and recorded in a headend logbook or entered digitally using a computer. A computer program that produces bar or histogram figures can be used to plot measured signal-level values for each channel. Such a display can show signal-level variations at a glance.

Some signal-level meters can produce a bar-graph display, and if the instrument has the data available on an output connector, a computer can be used to store the data and or display it on demand. Most headends have a TV set or a monitor, which should be used to examine picture quality for each of the carried channels. A headend technician should periodically check signal quality—at least three times each day. Any deterioration of picture quality can be determined and corrected before subscribers start to complain. Never forget that the picture is most perfect in the headend. Many systems have several trunk cascades connected to the headend directly. Each of these trunk cascades should have a headend test point available and a log kept of each input to the trunk run. Systems with several trunk runs connected directly to the headend are in an advantageous position when reverse plant activation is to be done.

Some systems have automatic test systems installed that measure and monitor various test parameters and communicate back through an upstream channel to the headend, giving an alarm as well as measurement data. A cable system may only monitor system power supplies or elect to monitor ending trunk amplifiers only. Online active system monitoring is the most elegant way to go even though it is expensive.

7.3 Headend and Hub Testing

Cable programming comes from four basic sources. The first is the television broadcast stations that are traditionally received off the air from tower-mounted antennas. The second is the satellite system cable operator's use of cable-only channels. The third is the locally generated cable channels. The fourth is a microwave link still used by some systems.

7.31 Off-Air Signals

The tower, lighting, and grounding system as well as the antennas and down-leads all require a maintenance program. This subject has been well discussed in a variety of sources that appear in the bibliography. Fortunately, some systems have technical personnel on their staff that can climb the tower and perform maintenance work. Periodically, light bulbs have to be replaced and antenna connections cleaned and remade. Also, the tower grounding should be tested and examined a couple of times per year. Guy wires may have to be tensioned to keep the tower straight. Tower snaking and tilt can be measured using a surveyor's optical transit or can be checked by a local civil engineer. Some cable systems make arrangements with local broadcasters to get the signal by optical fiber. Cable operators with an optical fiber trunk often have spare fibers going right by the television studio facilities. These broadcast television signals will be of studio quality for the cable operator.

7.32 Satellite Program Services

Other sources of signals usually consisting of the premium pay channels come from one of the satellite systems. Most cable systems have several parabolic antennas to pick up the various satellite channels for the cable-only channels. In most cases, these channels are scrambled or encoded, preventing unauthorized use. Cable systems at the headend have banks of receivers with built-in decoders called integrated receiver decoders (IRD). This equipment provides the cable operator with a base-band video and stereo audio signal that will be re-encoded with the cable operator's equipment and remodulated on a cable channel. The video quality tests are usually done on the base-band NTSC signal, according to the National Television Committee-7 (NTC-7) signal tests. This ensures that the NTSC base-band signal is up to broadcast standards. This procedure using a waveform monitor/vector scope appears in many of the references. The grounding of the satellite antennas is extremely important to protect the dish-mounted, sensitive solid-state, low-noise block down-converters (LNBC). The mounting frame should be connected to several ground rods around the base by heavy copper wire and bonded to the headend power ground. Also, keeping snow and debris from collecting in the dish is a must. Pointing this antenna on a satellite is discussed in Example 3-1 in Chapter 3.

7.33 Microwave Connections

Some cable systems still use microwave signal sources containing their cable programming. Microwave systems are discussed in Chapter 3. Baseband signal testing should be done on these microwave-received channels to ensure broadcast-quality signals are carried on the cable system.

7.34 Locally Generated Signals

The locally generated signals are signals only cable subscribers can receive. In the past, the local origination channels were often considered a burdensome and unnecessary expense. Other cable operators used these channels to carry interesting and informative local programs, thus promoting the cable system. A local studio means that cameras, switching equipment, and video and audio equipment have to be installed in a proper room. Personnel trained in television program production are necessary to provide quality programming. Again, the picture quality has to be closely monitored and held to NTC-7 standards.

7.4 Fiber-Optic Plant

Since many systems have or will be installing fiber-optic plant, it is timely to study these methods. A system may not have to install all of its own fiber, because the telephone companies, power companies, and independent common carrier systems have installed huge amounts of optical fiber and may have some fibers available for lease. Fiber-optic theory is available from sources mentioned in the bibliography, but from a practical sense, it will be taken up here.

Essentially, an optical fiber is a waveguide for extremely high frequencies (100 THz) or shortwave lengths (850–1550 nm). These wavelengths are below the visual spectrum. Therefore, because humans cannot see this light, optical power at sufficient levels causes serious eye damage. People working around such equipment should wear protective goggles. Also, one has to remember that optical fiber is glass, can shatter, and is sharp. Observing proper safety procedures when working with bare or stripped optical fiber is very important. Small pieces of broken optical fiber can get under a person's skin, causing discomfort and possible infection.

7.41 Fiber Cable

Fiber-optic systems are characterized as multimode or single mode. Multimode breaks down into either step-index or graded-index optical fiber. Step-index multimode fiber is the first generation, where the cladding (outside) glass has a lower refractive index than the core glass. This refractive index goes from a low to high value in a step change. As discussed in the references, light rays travel in many modes (paths) through the fiber, arriving at the end at different times. Such fibers demonstrate pulse stretching, thus limiting the bit rate. To fix this, manufacturers can produce an optical fiber that has the refractive index to decrease gradually from the core to the outside glass, thus producing graded-index multimode fiber. Now, rays of optical energy entering the core can speed up in the lower refractive index of the cladding, arriving at the end more in time with rays that traveled through the core and are slower at the shorter path.

Multimode graded index became the choice for local area networks (LAN) serving digital data communications. In general, multimode fiber is easier to manufacture because it has a larger core and thinner cladding; hence, it is easier to get light energy into a larger diameter core. Lower cost light-emitting diodes (LED) can be used for the transmitter and photo diodes used as the receiver. Multimode fiber systems have limited bandwidth and digital pulse rates. Single-mode fiber is used by cable operators to replace the coaxial trunk system, and is discussed in Chapter 4.

7.411 Fiber-optic transmitters for single-mode operations are available to cable operators by several manufacturers. General Instrument, Scientific Atlanta, C-COR, to name a few, all offer an excellent selection of equipment. These as well as other manufacturers are familiar with the cable industry and provide the interface equipment needed to go from optical fiber to the RF distribution plant. Usually, cable operators take a group of RF channels and directly modulate a laser-operated optical transmitter. Remember that the laser spreads its power level over a large band, depending on the size of the bandwidth of RF signals. Therefore, for longer distances, the RF band can be split into two or three bands per transmitter. Then two or more receivers can have their outputs combined. In most instances, high-power transmitters, transmitting through the normal fiber-optic distances, can carry the whole RF programming band. At one time, it was thought that using FM would be better, but technology in AM laser operation improved to the point that it became the method of choice.

7.412 The optical receivers used by the cable television industry are the integrated receiver to RF converter type. The optical cable is often terminated into the housing, and the optical signal is converted to RF at an appropriate signal level, comparative to a normal bridge amplifier level. This signal, as discussed in Chapter 4, is distributed to taps by using a five-amplifier RF cascade. Such an optical system from the headend to a receiving point is shown in Figure 7-14.

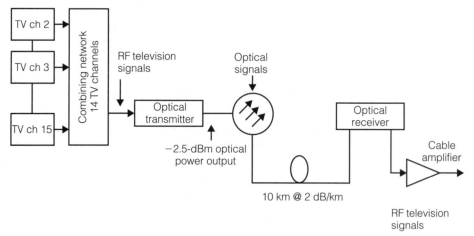

Power Budget

Cable loss over 10 km: 20 dB
Connector and splice loss: 2 dB
Optical power at receiver input: −2.5 dBm −22 dB =−24.5 dBm
if the receiver has a sensitivity of −30 dBm, the margin is:
−24.5 dBm −(−30 dBm) =5.5 dB, which is adequate.

Figure 7-14
AM fiber-optic transmission of cable television carriers

7.42 Fiber-Optical Testing

For many technicians, testing optical cable and devices is new and threatening. However, it quickly becomes apparent that the optical signal level is measured in terms of power in the quite normal unit of optical dBm. Optical power meters have been developed that are small and easy to use. Also, the optical time domain reflectometer (TDR) has been developed, which can be used to test optical cable.

7.421 Optical signal levels can be measured using an optical power meter that measures the level directly in dBm. Power meters operate at only one or two wavelengths, depending on whether the instrument is a single- or dual-wavelength type. The instrument range is large enough to be connected via an optical jumper directly to the transmitter. Reading this level as the input level and testing the level at the receiving point can yield the optical path loss. This method is shown in Figure 7-15. The optical level is displayed on a LCD display in numbers of dBm of optical power. The 1-mW level corresponds to 0 dBm.

These instruments are hand held, battery operated, and usually available with a variety of adapters for different optical connectors. Also, this instrument is quite rugged and relatively inexpensive, usually about $500 to $1,500. One manufacturer offers a power meter/light source combination with which a technician can use the light source with voice communications through a spare fiber to an accomplice with the same equipment

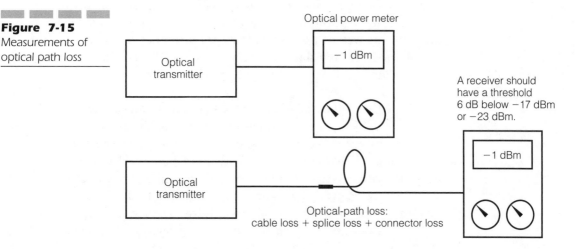

Figure 7-15
Measurements of optical path loss

Optical power meter

−1 dBm

Optical transmitter

A receiver should have a threshold 6 dB below −17 dBm or −23 dBm.

−1 dBm

Optical transmitter

Optical-path loss:
cable loss + splice loss + connector loss

at the opposite cable end. As a team, the two technicians can test the other fibers in the cable.

7.422 The optical TDR is an important and valuable piece of equipment and every system with any optical cable should have at least one. The typical price range for these devices is $6,000 to $16,000, depending on options. These instruments have large LCD screens displaying a loss profile of the fiber being tested. Cable systems often use cables carrying many fibers, where the various fibers can branch to other parts of the system at splice locations.

Optical cable should be tested before it is installed as well as after it is installed and before activation. The optical TDR can be used to establish the length of cable on a reel by measuring a fiber. The end point can be seen and identified, and any anomalies along the fiber can be seen on the display. Some cable operators do not measure every fiber, but most systems will measure at least one in each buffer tube for large fiber-count cables.

Optical cable can be either loose-tube or tight buffered-tube construction. Some cable systems choose loose-tube cable for aerial plant and tight buffered-tube cable for underground or where it is pulled through a conduit. Figure 7-16 shows how some faults are displayed on an optical TDR. Because testing of fibers is performed before and after installation, the test data has to be stored or filed for future reference. The TEK Ranger optical TDR has a keyboard option with which cable reel number, cable number, buffer tube identifier (usually a color), and the plastic coating color of the fiber can be entered, followed by the loss measurement. This instrument can store a large number of screen measurements that can be loaded to a computer for storage and or analysis. An optional disk drive is available as well as an RS 232 Port. Data can be stored and analyzed by Windows-supported software. Most optical TDR instruments are battery powered and built for troubleshooting in the field.

7.423 Other types of optical test equipment must have a high-level source of optical power that operates in the visible spectrum. Before connecting any optical cables to headend equipment, the cable can be tested by placing a light source to one end of a cable. Any bends in the optical cable that are too tight or too sharp will leak light and can be easily spotted visually. Many systems are concerned that optical cable will be cut, dug up, or damaged and repairs will have to be made. Single-mode optical fiber is spliced using a fusion-splicing device that encapsulates the splices in sealed PVC enclosures. More and more systems are electing to

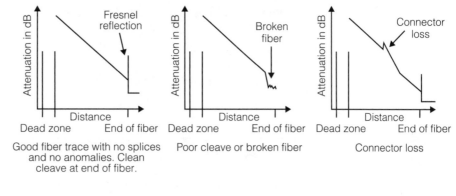

Figure 7-16
OTDR test configuration and traces of problems

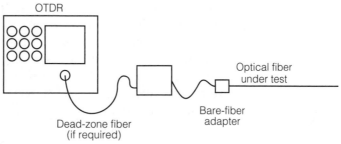

purchase a fusion-splicing device and to train several technicians to use it. These instruments are either very simple or microprocessor-controlled and vary in price from $8,000 to $30,000. Also, splicing has to be done in a relatively clean, dust- and dirt-free environment. Systems usually have a splice trailer or truck outfitted with commercial electrical power, heat, and or air conditioning. The optical cable ends are passed through a hatch door and placed on a brightly lighted workbench. When the splice has been completed and installed in the closure, it is passed back out the hatch for system placement. This procedure is quite precise but not difficult to learn. Most MSOs and larger systems have such equipment.

Quick fixes using mechanical splices are on the market but are difficult to install and get working. The best method for correcting a cut cable problem is to build in a redundant path for rerouting the optical signal into a node. This type of network topology has to be done during the design phase and initially built into the system. If the signal is lost on the primary fiber, a loss-of-signal switch can select the alternate fiber route.

7.424 Because the optical plant is designed to carry the RF television-modulated carriers, testing in the RF domain at the sending or transmission end will establish signal integrity at the fiber cable input. This signal should again be measured at the receiver output for comparison at the input. This type of measurement can be facilitated using one of the present spectrum analyzers that can drive a printer directly. Comparing a printout of the frequency spectrum of the carriers at each point is a quick way to establish network signal transparency.

7.5 Digital Signal Testing

The presence of digital television signals in a cable system requires more technical training. The broadcast television industry is involved with the conversion to digital signal transmission using 8-VSB of modulation. Cable systems plan to use quadrature amplitude modulation (QAM) mainly because it will work with consecutive channel operations. Therefore, it is apparent as of this writing that cable systems will elect to convert these signals from 8-VSB to QAM-64. Digital signals will have to be tested in the digital domain and in the RF-modulated domain to ensure quality transmission.

7.51 Instruments for Digital Testing

For many years, digital signals have been used in the telephone and the computer industry. Several instruments have been developed for testing digital signals. The modern oscilloscope—with digital sampling, storage, and display—is probably the singular most important piece of test gear. The logic analyzer and signature analyzer can test for digital signal errors by recognizing and testing the accumulated errors over a time period. Special test equipment used to analyze specific digital signals for telephone systems has been used for several years to provide the bit error rate information. The cable television industry is responding to the need for instruments that are necessary to test the modulated QAM-64 signals.

7.511 As mentioned earlier, the signal-level meter (SLM) is one of the most important and most used instruments in the cable television industry. SLMs needed to test digital modulated carriers are on the market now with more exotic types just recently introduced. Acterna, Wandel and Goltermann, Sadelco, Trilicthic, and Sunrise Telecom are some of the manufacturers that offer digitally modulated carrier-level testing. Digital signals as a base-band pulse train are measured by using a high-quality, high-speed oscilloscope set to produce an "eye" diagram, named so because the oscilloscope trace resembles a human eye, as shown in Figure 7-17. Examining this trace can produce information on the quality of the digital signal before modulation. Ideally, the intersymbol interference should be 0 dB. Digital signals can be examined and measured using a high speed, high frequency oscilloscope. Digital oscilloscopes offer digital memory and trace storage, as well as wideband, high-speed operations.

7.512 For NTSC television VSB modulation, a negative carrier modulation is specified. This conserves transmitter power, heating, and so on. In short, it is easier on the transmitter system. This type of signal has the

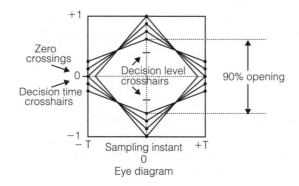

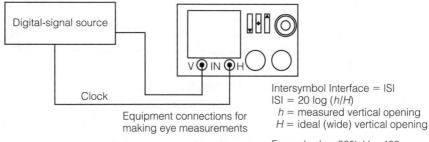

Intersymbol Interface = ISI
ISI = 20 log (h/H)
  h = measured vertical opening
  H = ideal (wide) vertical opening

Example: h = 90% H = 100
ISI = 20 log (90/100) = 0.915-dB ISI degrade

Figure 7-17
Eye diagram of a digital signal

peak RF level at the video horizontal sync pulse that occurs every video line for about five microseconds. The SLM has to have a peak detector, because this peak signal level is what drives the amplifier cascade to maximum. For digital signals, an averaging detector is more appropriate. Some of the future SLMs will be able to measure NTSC signals with a peak detector and digitally modulated carriers with an averaging detector. Present-day SLMs offer more options, are hand held, have a longer battery life, and provide a larger scanning LCD screen readout. Some have leakage/ingress options as well as memory for recording measurements. Some of these instruments coming on the market will be able to give a digital signal quality check by plotting a constellation diagram of QAM signals. As discussed previously in Chapter 5 and illustrated in Figure 5-32, the constellation diagram is a pattern of dots placed in each of four quadrants. Each dot location in a QAM-16 represents a pattern of four dots in each of the four quadrants for a total of 16 dots. Each dot represents a number in binary code 0000-1111 (0–15 decimal), or four bits per dot. Each dot also represents an amplitude and a phase, hence QAM.

An important new instrument for digital television testing has been produced. This hybrid type of instrument can display a QAM digitally modulated television signal in a variety of modes, such as a 6-MHz spectrum (level/amplitude), average signal level, a histogram display (bar graph), and a constellation diagram in both one quadrant or all four quadrants (zoom mode). All screen displays are on a well-sized LCD display and properly annotated. The clarity and tightness of the dots indicate a good signal-to-noise ratio and is related to bit error rate. This instrument can function in the standard NTSC mode, measuring both the audio and visual carrier levels in dBmV, as shown in Figure 7-18. A digitally plotted spectrum of NTSC television channel 3 is shown in Figure 7-19.

Measuring QAM digitally modulated signals with the Sunrise Telecom instrument is easy using the simplified keypad and the LCD screen display. QAM digital signal levels can be measured and displayed in simplified form, as shown in Figure 7-20. Also, the 6 MHz-wide band, indicating a QAM-64 signal, is shown as a spectrum in Figure 7-21.

The constellation plot provided by the instrument has a lot to say about the digital signal quality. The digital signal displayed is an accumulation of measurements according to the phase and amplitude positions of the QAM signal. These measurements are displayed as dots on the constellation diagram. If the accumulation of the dots is tight and small, the signal quality is excellent. Figure 7-22 shows channel 109 with all four quadrants plotted, and Figure 7-23 shows essentially the same signal with zoom activated, showing just the upper right-hand quadrant.

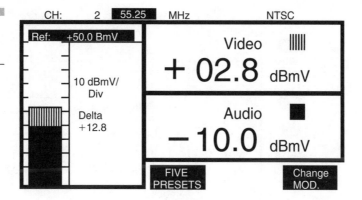

Figure 7-18
Video/audio
carrier-level display

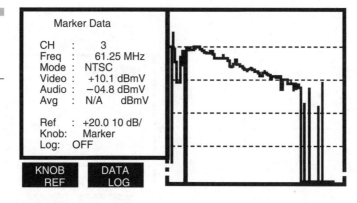

Figure 7-19
Digitally plotted
display of a TV
channel 3 spectrum

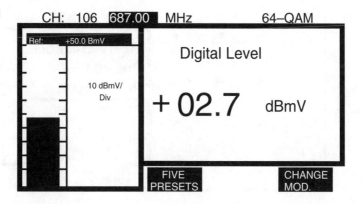

Figure 7-20
Display of QAM-64
digital signal level

Figure 7-21

Spectrum of QAM-64 signal

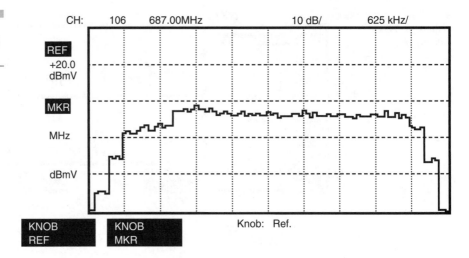

Figure 7-22

Constellation display of QAM-64 signal on TV channel 109

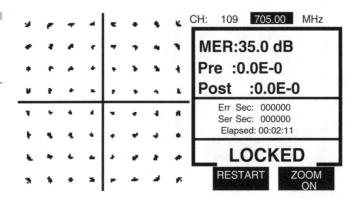

Figure 7-23

One quadrant of constellation display of QAM-64 signal on TV channel 109

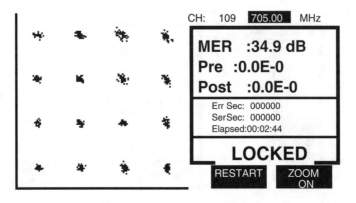

The alphanumeric readout indicates MER −35.0 dB, which means modulation error ratio (MER), and is related to signal-to-noise ratio. This is measured by comparing dots on the constellation diagram to the position they are supposed to be. This essentially is a statistical analysis result of the dot location variations. The Pre 3.1E-3 notation is a measure of uncorrelated data in scientific notation. For example, 3.1E-3 is the same as $3 \times 1 \times 10^{-0.3}$, mathematically. The Post notation is given after forward error correction (FEC) has been applied and shows the signal improvement in the same notation as the pretest. The MER, Pre, and Post are related to overall bit error rate (BER) but indicates more about the signal errors and the causes.

Once operators get used to this type of instrument to measure QAM signals, these features will be greatly appreciated. As of now, only high-speed oscilloscopes and sophisticated spectrum analyzers are available for measuring QAM digital television signals. No doubt, other test equipment manufacturers will offer such equipment, but Sunrise Telecom offers the CM 1000 QAM monitor/SLM that comes with a battery-operated, weatherproof case and is reasonably priced. Personnel trained in its use will find it extremely useful in maintaining QAM modulated digital television signals. The screens in Figure 7-22 and 7-23 show QAM-64, the digital format of choice for cable systems offering digital television service.

7.513 Many cable systems operating in the larger urban areas will no doubt find that some sources of television programming will be available from optical-fiber systems operating with SONET encoding. Now the job at hand is to extract from the synchronous payload envelope (SPE) the data packets containing the programming information. Also, SONET terminal equipment will have to be installed in the headend. Technical personnel will have to be trained in the SONET protocol as well as the techniques necessary to test the quality of the signal and point to any trouble spots, should they occur.

Because this instrument contains both a transmission and receive system it can be used in end-to-end testing and daisy-chain test loops, to name a few applications. Such equipment with its options can be somewhat expensive but will often be necessary for systems connected to SONET facilities. With the interconnections to other communication carriers, many cable television systems will be and are players in an advanced cable telecommunication industry. In the foreseeable future, fiber to the home (FTTH) will be a reality, with fiber to industry and commercial enterprises occurring first. All telecommunication systems will be carried on optical fiber delivered to and from users. Technicians and engineers will have to get training on optical-fiber technology as well as high-speed digital signal-testing methods. System maps, test data, and maintenance records should be stored in digital files for fast retrieval and analysis.

7.6 Typical System Problems and Solutions

Cable communication systems have wide and varied problems. Most cable television systems carry a.c. voltage to power amplifiers and fiber-optical nodes. The telephone industry carried −48 v d.c. and 90 v.a.c. at 20 Hz (ring power) for years. Typically cable system optical nodes are powered from the coaxial cable plant. Cable systems in the beginning used 30 v.a.c. power supplies, oftentimes pole mounted, that used the ferroresonant method of voltage regulation. As amplifier cascades grew longer, the required voltage was increased to 60 v.a.c. with some at 75 v.a.c. Many of today's systems use 90 v.a.c., and if required these power supplies are powered by rechargeable batteries for standby use. When a commercial power failure occurs, the standby battery pack drives an inverter circuit providing the necessary voltage of 90 v.a.c. Proper maintenance procedures are necessary to test and cycle these standby power supplies. These power supplies increase the system reliability by several orders of magnitude.

7.61 System Powering and Power Supplies

For many years, coaxial cable systems have carried their own power for operating cascades of amplifiers. Power supply voltages have increased from 30 v.a.c. to 90 v.a.c. over the years. The most common voltage was 60 v, which increased to 72–75 v, then to 90 v. Most power supplies used a ferroresonant transformer and were a non-standby type. The ferroresonant transformer provided a regulated alternating power supply. A large capacitor was connected across a separate transformer's secondary winding to form the resonant circuit. A circuit diagram of this type of cable power supply is shown in Figure 7-24. Voltage regulation was usually on the order of less than 5 percent for most supplies of this type, which was a workhorse for the industry. Many small or rural systems may still have some in service.

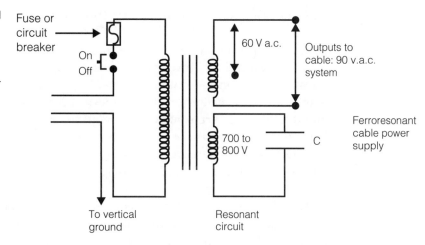

Figure 7-24
Ferroresonant power supply of 60 to 90 volts a.c.

Many systems have replaced such power supplies with standby types. Basically, there are two types of standby power supplies: the switched supply and the uninterruptible supply. The switched supply turns to standby battery power upon commercial power failure. When commercial power returns, the unit switches off and places the battery bank on charge.

Early units used relay-type switching with a switching time on the order of a quarter of a second. Later generation supplies using solid-state switching usually switched in a few milliseconds. Standby times were usually two to four hours, depending on how many batteries were used.

Models with various standby times were available from several manufacturers. Some early supplies used the ferroresonant transformer as a basis. The switch would disconnect the commercial power from the transformer primary and connect a solid-state 60-Hz inverter powered by the battery bank to the transformer primary. It became evident that the core saturation current caused too much battery drain, which decreased standby time.

Some manufacturers used two separate transformer types, the ferroresonant for commercial power and an ordinary step down for the standby operation. In standby mode, solid-state regulators provided the regulated supply voltage. Unfortunately, these supplies, with two transformers and the batteries, were extremely heavy and difficult to pole mount. Thankfully, the uninterruptible supply, using a ferroresonant transformer with solid-state regulators and controls, is the mainstay of many upgraded/rebuilt cable systems with active upstream and downstream capacity. Efficient design gives standby times for up to two hours using two batteries. However, most units have three or four batteries for extended "on" times.

Several present-day designs use modular construction, in which the electronic control and inverter systems can be housed in various cabinet sizes that can contain different battery configurations. Front-panel test points, an LCD readout of status and test information, breaker switch, and dual outputs are some of the main features of these devices. Problems associated with these power supplies can be devastating to a system because they were installed to solve the power outage problem and increase signal reliability. Therefore, proper installation and maintenance are extremely important.

As with any pole-mounted equipment, the weather and seasonal changes provide a difficult working environment. Temperatures can be as great as 100° F, with humidity ranging typically from 40 to 100 percent. Effects of lightning and associated power surges also have to be absorbed by the cable power system. Present-day power supplies used by most of the major systems use 90 v.a.c. to power the plant with a maximum volt-ampere rating of 1800 va. Some of these supplies offer dual voltages and can be used to supply 60 v.a.c. to 75 v.a.c., up to 90 v.a.c. Because keeping these power supplies in top-notch condition is extremely important for signal reliability, a full-time, properly trained maintenance crew is required to visit each supply and perform maintenance checks on a periodic basis. Batteries should be checked for any leakage, corrosion at the terminals, and any collection of foreign material in the cabinet. Electrical measurement data for each supply should be entered into the plant maintenance records.

7.611 Problems with cable system powering breakdown into short circuits or open circuits is that it causes an outage to occur. Troubleshooting the sources of short or open circuits can be time-consuming, particularly if the area affected is large. Driving back and forth making measurements is the "fencing with windmills" syndrome. Systems using status monitoring on their power supplies usually can solve these problems quickly. Some systems elect to place status monitoring only on the power supplies. Data taken from the power supply's status monitors can be scanned at the maintenance office and the problem can often be pinpointed. Parameters from the status monitor can determine if commercial power is present or whether the supply is in standby or normal mode, battery voltages, current draw, and of course, open or tripped circuit breakers. Usually, the first notification of a problem is a signal outage reported by a subscriber by telephone.

The proper methodology is to dispatch a technical repair crew to the general area of the problem while recording calls from the affected area. The repair crews should have available copies of system maps, either on paper or on disk, which can be shown on a laptop computer monitor in the repair truck. If the affected area is confined to a system power supply, then the repair crew should check the status of the unit. If the supply has an ammeter or an LCD display showing the current drawn from the supply, an excess of current can indicate a problem. If the indication of current is changing, then an open or short circuit in one of the system branches could be causing the current variations. Figure 7-25 illustrates this sort of problem.

As shown in the figure, the current provided by the supply flows to the power inserter, where it splits in both directions. Ideally, equal currents as calculated by the design parameters flow in both directions. In many cases due to plant additions, this equal current split is destroyed. In order to isolate each current's direction, an ammeter can be clipped across one of the fuses. Now the fuse can be removed so the a.c. ammeter can read this current. This procedure is shown in Figure 7-26.

Now the high current indication on the ammeter identifies the cable section with a short-circuit condition. Some manufacturers still offer the self-resetting glass-enclosed circuit breakers, which can be connected through

Figure 7-25
Current distribution
from power supply

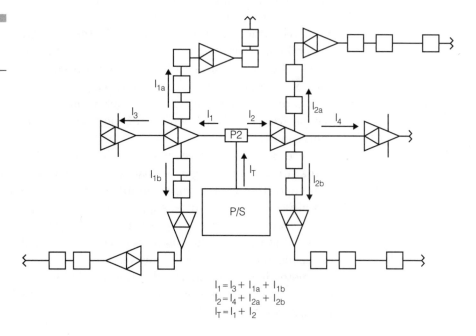

$$I_1 = I_3 + I_{1a} + I_{1b}$$
$$I_2 = I_4 + I_{2a} + I_{2b}$$
$$I_T = I_1 + I_2$$

Figure 7-26
Ammeter positioned
to measure current.
To measure, remove
fuse, read meter, and
replace fuse

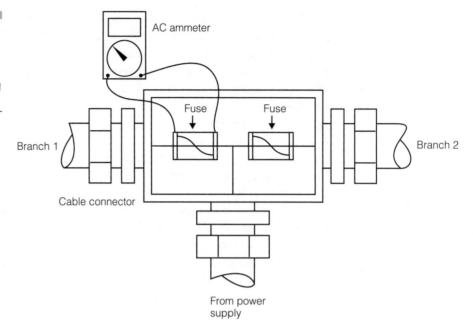

Figure 7-26
Ammeter positioned to measure current. To measure, remove fuse, read meter, and replace fuse

clip leads across the fuse holder in the power inserter. By observing the contact positions through the glass envelope, a technician can identify the high current leg of the cable system. This method is illustrated in Figure 7-27. Once the portion of the cable system with either the open- or short-circuit condition has been identified, the task is to track down the cause of the problem. Usually, the area to next check out is where the subsequent cable split occurs. Either a line-splitter or a directional coupler will be at this location and usually the output ports are fused. Substituting either an a.c. ammeter or glass envelope circuit breaker will identify the cable section with either an open circuit (no current) or an overload (high current) condition.

Since power to the feeder or distribution system is derived from the trunk through the bridger, identifying the defective cable section requires some detective work. Trunk amplifier bridger ports are fused and also contain some form of surge suppression. Such surge suppression devices range from gas surge suppressors to the metal oxide varister (MOV) types. Testing of the fuse is performed by setting the DMM in a.c. voltage mode and measuring the a.c. voltage across the fuse terminals. When the meter indicates full voltage, the fuse is blown, identifying the high-current cable section. Conversely, zero voltage means the fuse is okay and passing current.

Figure 7-27
Self-resetting circuit
breaker

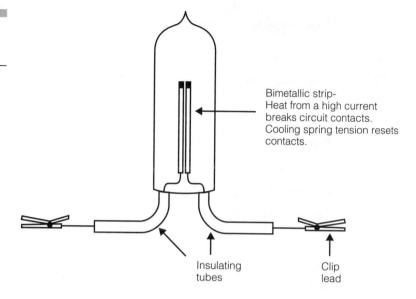

Bimetallic strip-
Heat from a high current
breaks circuit contacts.
Cooling spring tension resets
contacts.

Insulating
tubes

Clip
lead

Tracking down power distribution problems can be frustrating and time-consuming. In many cases, a cable system plant starts out properly designed and constructed. However, plant additions and adjustments that normally progress in a maturing plant can upset the power distribution balance. This results in different currents in the two parts of the power inserter device. Tracking down these open- and short-circuit conditions in the cable distribution system can be laborious. The causes of open and short circuits are given in Table 7-3.

Solving open- or low-current conditions and short- or high-current conditions are types of power supply problems that can be aided by a good record of the correct values of the currents. On-site, current measurements compared to the correct values will point the way to the cause of the problem.

7.612 The effects of lightning can cause many problems for the power distribution system. Nearby lightning strikes cause large electric and magnetic fields, which induce large voltage spikes into the commercial electric plant. These large electric and magnetic fields induce voltages into any conductor suspended in free space. Since most of these conductors are grounded, the currents caused by induction are passed to ground. Excess currents can cause heating of connection points, resulting in visible charring. The lightning damage caused by severe strikes can be inspected and repaired. Power surges in the electric plant can also cause power supply

Table 7-3

Common Causes of
Open- and Short-
Circuit Conditions

Causes of Open Circuits	Causes of Short Circuits
Connector pull-out	Connectors (pieces of metal scraps)
Loose seizure screws	Capacitors, short-circuited due to lightning or power surges
Blown fuses	Crushed cables
Corrosion in connectors	Amplifier power supply problems
Amplifier housings	Vandalism of power supplies
Passive components	
Tripped circuit breakers	
Vandalism of power supplies	
Cut cables	
Effects of lightning and power surges	

circuit breakers and/or fuses to blow, but they can be protected by surge protectors installed in the amplifier housings. Voltage surges that are not corrected by the power supply surge protectors can cause fuses to blow in the cable directional couplers and signal splitters.

In some cases, the subscriber taps will become defective after nearby lightning strikes. Some of the ceramic capacitors in the taps will have their working voltage greatly exceeded due to a voltage surge and burnout. For aerial plant, taps with a short-circuited capacitor will have a greatly reduced output port signal level. A line technician can remove the seizure screw cover and smell the burned-out capacitor, thus identifying the defective tap. This is truly "sniffing out trouble." Often, a whole string of taps on a feeder branch can become defective and will have to be replaced. Because most taps have a circuit board fastened to the bottom plate, the change-out procedure does not require a splicing technician. Most of the time, inspection of the components on the circuit board will indicate the burned-out capacitor. Most cable systems do not replace the capacitors because other problems with the tap plate may surface later. The distribution amplifier, or line extender amplifier feeding the affected cable run, should be checked for visible damage and for proper input/output signal level. Unfortunately, lightning problems may take time to surface. This can make troubleshooting the affected area and pinpointing the faulty devices difficult.

7.62 Standby Power Supplies

Standby power supplies in cable systems add to the list of maintenance problems. Batteries have come a long way in improvements, however, and several types are available. The battery recommended by the manufacturer is usually the best one to use. Early standby power supplies used heavy-duty lead acid storage batteries similar to those used in cars and trucks. Carefully charging such batteries is critical so as not to overcharge and cause leakage of gases and the electrolyte. Many standby units presently available have proper charge controllers as well as solid-state inverters. Still, these units should be periodically inspected and tested.

Standby power supply tests are mechanical or electrical. Electrical tests include current and voltage measurements taken in the normal power mode and in the standby mode. When in the standby mode, the battery voltage should be monitored for 10 to 15 minutes, which will indicate any significant voltage drop due to battery deterioration. Technicians making these measurements should use a digital multimeter for the a.c. and d.c. voltage measurements. A clip-on ammeter is useful in measuring the a.c. current from the commercial power source and/or the inverter to the ferroresonant transformer.

Visual inspection could determine that loose components have been caused by normal vibration or possible vandalism. The most common of the visual problems is battery leakage and corrosion. Such corrosion should be cleaned and any acid or electrolyte should be neutralized with baking soda or an alkali spray. After cleaning, a silicone spray protection should be applied to seal against any action due to moisture. Standby power supply inspections should be made on a monthly or bimonthly basis, and many system operators have a crew of technicians trained specifically for this work. Documentation for these inspections should be recorded on paper or electronically on a computer disk and filed in a power supply maintenance log. An example of such a form is shown in Figure 7-28.

Probably the most catastrophic power supply problem is if an automobile strikes a utility pole with a standby supply mounted on it or if a pedestal containing a standby supply is hit. In either case an outage usually occurs, and if personal injury results, repairs can only be made after the accident scene is cleared. If at all possible, a non-standby unit operated by a vehicle generator set can often be jumped into the circuit with clip leads to get the system temporarily back into service. As it is often said, "An ounce of prevention is worth a pound of cure." Therefore, the installation site for standby power supplies should be carefully chosen to be in as safe

Figure 7-28

Log form for standby power supply

STANDBY POWER SUPPLY LOG_____

NO_____   TOWN_____ ST._____ PED POLE#_____

DATE/TECH	TEMP (F)		SEQENCE	AC VOLTS	NORM PWR. LT.	NORM SUB. LT.	INV. RDY. LT.	CHARGE LT.	LOW BAT LT.	BATT VOLTS DC	REMARKS
		NORM									
		STBY									
		8 MIN.									
		15 MIN.									
		NORM									
		NORM									
		STBY									
		8 MIN.									
		15 MIN.									
		NORM									
		NORM									
		STBY									
		8 MIN.									
		15 MIN.									
		NORM									
		NORM									
		STBY									
		8 MIN.									
		15 MIN.									
		NORM									

a place as possible. Corner poles and street intersections should generally be avoided. Pedestal-mounted units should be placed as far off the road as practical and should be marked by a warning stake.

Maintaining the batteries used in standby power supplies is a major job. Often, batteries have to be replaced, the result of the periodic testing program. New batteries should be charged and load tested before deployment in the system. The battery maintenance area should be well ventilated so any out-gassing during charging can be dispersed. Storage batteries or gel cells contain acid as the electrolyte, which can cause serious skin burns on a person's skin. A supply of baking soda or a solution of baking soda and water should be kept nearby in case of an acid spill.

Also in the battery room, a charger and a load test should be installed to service any batteries requiring maintenance. Often, one or two standby units can be used as a charger because in most instances charging is unnecessary. Switching these supplies into standby mode and timing the battery voltage and current can test the capability of the battery to maintain the load. Systems requiring testing of a large number of batteries may wish to construct a battery charger, as shown in Figure 7-29.

The most common problems occurring in standby power supplies involve the battery system. Either the batteries fail or the charger fails to keep them fully charged after an outage. High internal temperature due

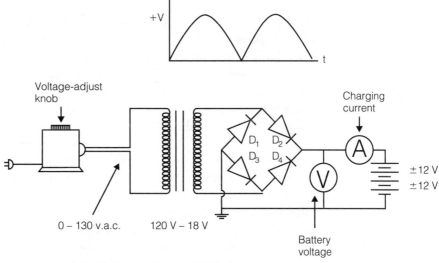

Figure 7-29
Pulse-type battery for charging standby batteries

D_1 D_2 D_3 D_4 are 10-amp, 50-V diodes

Charger charges two 12-volt standby power supply batteries in series. Adjust input voltage for desired charging current.

to ordinary heating during operation, along with a high summer ambient temperature, can shorten the battery life. When the power supply is operating in the non-standby mode, the ferro transformer is generating heat. Charging the batteries after an outage also causes heat, worsening the problem.

Many standby power units contain a cooling fan, which can effectively remove heat. Maintenance inspections of standby power units should include inspecting the fan system, which may include a thermostatically controlled switch. The internal battery charger should be cycled and checked to make certain the trickle charge current to charge battery is not too high, which could overcharge the batteries. Most standby units have the battery compartment in the bottom shelf of the cabinet, preventing the transformer heat from rising into the battery compartment. As a rule of thumb, the battery-charging parameters as given by the manufacturer should be followed when setting the standby power unit's charging controls. It is easier for technicians servicing pole-mounted standby power supplies to use a hydraulic lift truck.

7.63 Coaxial Cable System Problems

As cable systems age, the cable systems trunk and distribution systems change along with the size of the subscriber community. More cable, passives, amplifiers, and power supplies are added to plants to accommodate expansion into subdivisions and residential developments. As the plant expands, it sometimes bears little resemblance to the original build. Unfortunately, the newer service areas are connected to the system by the old and problematical cable system.

Cable systems finding themselves in this position—faced with advanced digital television, video on demand, the Internet, and other additional services—must adapt to these new technologies. Some cable systems with low debt and good positive cash flow have elected to rebuild their systems. Other systems have decided to either sell out or simply do nothing. Companies buying such systems in most cases rebuilt them. The companies that chose to do nothing continue to operate with the same programs.

7.631 Connectors, directional couplers, and splitters pass the signal as well as system power. Any loose or poorly made connection-conducting current will heat and eventually cause an outage. Connectors, like any other cable system components, have improved over the years.

As stated earlier, the older parts of a system feed both the newest- and oldest-constructed parts of the system. Any problem occurring in the older parts can affect a large number of subscribers. Leakage testing, as required by the FCC, has caused many cable operators to focus on old and defective connectors. A common connector problem is often called a pullout; the cable has essentially pulled out of the connector enough to break system power. Pullout can cause intermittent power and signal variations. Such problems are usually the most evasive and difficult to solve.

Also associated with connectors is the cable itself. Flexing of the aerial plant due to wind, rain, snow, and ice causes strain on the cable, usually at the connector. This strain can cause the aluminum cable sheath to crack open, allowing water and moisture to enter the cable dielectric. Moisture and electricity react to form extreme corrosion and system failure. Most systems use heat-shrink sleeves on the connectors that protect against this. Cable systems in a coastal environment usually have jacketed cable with heat-shrink tubing on the connectors, which adds significantly to the life of the cable plant.

Early manufacturers of cable passives paid little attention to the effect different metals had when in contact with one another, particularly with added moisture. Cable systems along the coast often watched their taps and passives literally disintegrate before their eyes. Present-day manufacturers, for the most part, are well aware of the metallurgy involved and make high-quality devices oftentimes coated and with protective O-rings and seals.

Construction practices have improved drastically as well. Semicircular or rounded cable expansion loops have been replaced by the flat-bottomed type made by a ratcheting cable bending tool. This type is the most accepted and most used by cable systems today. Proper construction practices and methods are discussed in several of the references listed in the bibliography.

7.632 The failure of amplifiers in the system amplifier cascades can present myriad problems. Never forget that amplifiers early in the cascade can cause serious problems at the ends. Most systems, at the completion of construction, go through the initial balance followed by sweeping, final balance, and proof-of-performance testing. At this point, the system will never be in better shape. Now, the shake down cruise begins. The leakage-testing program identifies loose or poorly installed connectors and devices, as well as loose amplifier housings. As a rule of thumb, after the initial turn-on and balance, the system should be leakage tested before sweeping and final balance takes place. This ensures that final balancing and sweep-response testing does not compensate for a leaking and faulty cable plant.

Larger cable systems usually maintain an in-house amplifier laboratory where an amplifier can be bench-operated, aligned, and adjusted. Amplifiers brought in from the outside plant can be tested to ensure that they are indeed defective before sending them to the manufacturers. Some systems maintain an inventory of components and integrated circuits, essentially performing their own amplifier repairs down to component level, but do not elect to exchange modules, sending the defective ones out to a contract repair facility. Some systems have an amplifier housing of each type that is used in the system setup on a wall with a bench in front where a module can be tested.

When an amplifier module fails in the plant, a replacement can be pretested and set up, thus speeding the repair process. Such a test setup enables the technical department to test modules so they can be tagged with various operating problems before sending them to a repair facility. This same setup can be used to ascertain whether repaired units operate according to specifications. Large systems often have their own repair facilities that support all their systems in the surrounding area.

Common amplifier problems are often related to the switching power supply module. This type of power supply is well regulated as per line power and load power. This type of supply rectifies the 60 or 90 v.a.c. cable power and then converts this d.c. power to a higher frequency a.c. that is regulated precisely and then converted down to the standard d.c. voltage used by the amplifiers. The transformer that controls this high-frequency a.c. voltage emits sound at the a.c. frequency. Normal operation sounds are different from an abnormal operation, and often an experienced technician can tell which is normal. Once suspected, a voltage measurement will often confirm the faulty unit. Capacitors of the electrolytic type are the most likely to fail in a power supply. Also, diodes on the input rectifier are known to often fail after a thunderstorm. Surge suppressors should be inspected and, if suspected, they should be replaced.

7.633 As in the downstream system, the reverse system can have similar problems, but usually not as many subscribers call the office. A monitoring system, either manual or a status-monitoring system, will indicate a problem and usually identify the general problem area. In many instances, a technician at the headend or hub will detect an abnormal condition on the reverse signal and immediately start to explore the cause by further testing. Some systems power the reverse system from the same units feeding the forward. Therefore, failure of a power supply can affect both forward and return systems. Such a condition will usually point to the faulty power supply.

Intermittent problems are the most difficult to solve, but problems of this nature with the reverse system can at least be monitored at the headend. The repair crews can go into the system, inject signals, and make adjustments and tests while keeping in touch with the headend or hub by two-way radio or cellular phone. Most systems that have an active return system will have test equipment for aligning and troubleshooting the system. As always, a high-quality calibrated spectrum analyzer in the hands of a properly trained technician will usually find the problem unit, so a replacement can be quickly made.

7.64 Fiber-Optical System Problems

Systems using fiber-optic technology usually have another set of problems. Luckily, some of these problems are easier to solve simply because there is little, if any, electronic equipment between transmitting and receiving point.

7.641 Monitoring laser currents at the optical transmitters will indicate laser operations. As lasers age, they tend to draw more current. If the laser in an optical transmitter starts to show an increase in current that is approaching an out-of-normal range, the optical transmitter should be changed out. Still, a laser in the transmitter can suddenly stop operating, causing an outage. If this occurs at the headend, a change out can be made quickly, because most systems have a headend technician on duty most of the time. Laser transmitters on the upstream or return system that fail have to be tracked down and replaced. Optical transmitters and receivers are generally quite rugged and long-lasting. Most optical electronic equipment contains lamp monitors or meters that indicate normal or faulty operating conditions, which are extremely helpful in identifying failed or failing equipment.

Present-day optical systems often have equipment test points where an optical power meter or other test equipment can be used to test system performance. Optical power meters can perform field tests for optical performance, but a complete terminal-to-terminal test using an RF spectrum analyzer can confirm system operation. As the optical plant is expanded to dense wavelength division multiplexing (DWDM), in which several optical carriers generated by lasers carry many more services, an optical spectrum analyzer can be very helpful. Optical spectrum analyzers essentially display the optical carriers on a CRT screen, shown in Figure 7-30, similar to that on an RF spectrum analyzer. The optical carriers can be tested for proper wavelength (or frequency) and optical dBm levels. Lasers not having proper output power can be easily identified and replaced before service is affected. Cable systems that are connected to a SONET link may want to opt for a SONET testing system.

7.642 Optical cable that is properly installed and spliced together should cause few problems. The most common problem often occurs at a terminal location where the main cable is terminated and connected to equipment using jumpers and pigtail optical cables. Kinking or sharp bends can cause a severe loss. Proper cable tracks and trays should be used to ensure that the cable is loosely routed. After installation and before activation, the spliced-in jumper cables can be tested with a light source operating in the visual spectrum and placed at the optical connector. Light will shine through the plastic jacket of the jumper cable if the bend is too tight, so the problem can be easily seen and corrected. Once this is accomplished, technicians working at or around these optical jumper cables should be careful not to cause any sharp bends in the cables.

Figure 7-30
Display from an
optical-spectrum
analyzer

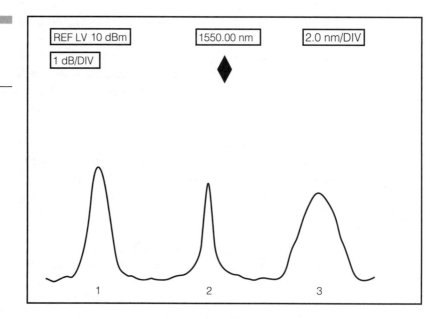

REF LV 10 dBm   1550.00 nm   2.0 nm/DIV

1 dB/DIV

1   2   3

3 optical signals: signal 1 — +5 dBm, 1540 nm
                       signal 2 — +4.3 dBm, 1548 nm
                       signal 3 — +4 dBm, 1556 nm
Signal 3 has a wider spectrum width.

Other problems with optical cables are often man-made. Cars and trucks hit poles, above-ground pedestals, and terminal units, thus causing smashed or broken cables and equipment. Rapid change-out is the only cure for such problems. Operating cable systems have also reported such problems as bullet holes in optical as well as coaxial cable plants. Fires in buildings next to aerial cable plants can damage the plants. The possibility of such problems should be addressed at the time of initial build, but changes in the plant due to extensions and system rerouting can lead to critical equipment being poorly placed. As mentioned earlier, power supplies should not be placed at street intersections. Also, an optical signal-mode pedestal should not be placed at a street intersection.

Most other problems with optical plants include keeping it clean and free of moisture. The most important point to remember is to clean the optical connector before reconnecting to a piece of equipment. An almost unseen (even with a magnifier) speck of dust can cause a significant decrease in signals.

7.65 Headend Problems and Maintenance

Headend problems can be, at times, elusive and confusing, particularly with the off-air television broadcast stations. Some cable systems use a microwave facility to connect remote programming to the hub/headend facility. In many instances, satellite Earth receiving stations provide a multitude of channel selections, making up the system offering. Often, this installation is done at the headend site. Each one of these signal sources seems to have some unique problems that have to be solved by the technical staff.

7.651 Local television broadcast stations are received by tower-mounted antennas appropriately aimed at the station location. The size and position on the tower depend on the required signal strength for the television station signal. Careful study and planning determine the proper antenna and any pre-amplification necessary to provide a high-quality television signal to the headend. Assuming that initial design has been adequately done and the system has been properly installed, typical and not-so-typical problems will be investigated. It must not be forgotten that what has been constructed is a tower-mounted listening post where the antenna gain and direction are peaked to the desired station's frequency. What happens between this facility and the station is beyond control of the cable operation. A problem involving a headend facility is given in the following case study:

A cable system headend detects an interfering signal, nearly obliterating a usually good channel 7 broadcast station, rendering it unwatchable. Because afternoon soap operas appearing on this station are coming on, telephone calls from subscribers start to come in. A technician at the headend is alerted and switches to an auxiliary search antenna. The signal is equally bad on the search antenna and is made worse slightly with a small change in direction. By looking at the detected video, a spectrum analyzer determines that the interfering signal is also a television signal.

The chief technician outfits a truck normally used for leakage detection and sets the dipole to channel 7, connecting a TV set and a signal-level meter to the antenna. The driver starts traveling toward the source of the interference, thinking that it might be caused by a severe cable leak. Since the signal is so strong, it seemed that if a cable leak caused it, other channels could be affected, and they were not. After circling around within a

1-mi radius from the headend, a strong co-channel 7 picture is received on the TV set and a +20-dBmV signal is recorded. In the immediate area is a residential home with a television antenna with the high-band flat lead broken from the high-band antenna section and drooping down past the antenna-mounted pre-amp. This length of twin lead (input to pre-amp) is drooping alongside the pre-amp down lead (output or pre-amp), causing a feedback loop. The flapping lead picks up enough channel 7 signal to produce a strong signal through the pre-amp oscillation. Knocking on the door gets no answer. A neighbor says that the person moved out a while ago. A pair of side-cutting pliers is applied to the down lead, thus removing the pre-amp power, and the problem is immediately resolved.

The case just described is a case beyond control of a cable operator. Clearly, this is not a case of malfunctioning cable equipment or a problem with the cable system, but it resulted in destroying one of the off-air stations. Because most people in the community were cable subscribers and removed their antennas, this problem did not affect them. Troubleshooting requires technical expertise and good detective work.

Other headend problems usually concern failed processors or modulators due to small contact corrosion problems. Cleaning with a contact cleaner often restores operation. Most headend equipment is so overdesigned that units operate for years without being touched.

Tower-mounted equipment, on the other hand, usually does have failures. Antennas with wiring harness TY-wrap are known to have problems. First, some TY-wraps break, allowing the harness to hang down and swing in the wind. This causes stress at the connection point, allowing the lead to break from the antenna. Clearly, this is a problem that, if noticed, can be fixed. Usually, the tower has to be climbed and the wire secured before it breaks. If only one element is bad, it can be lowered to the ground, repaired, and put back up. Pre-amps and tower-mounted down converters after several years of service also break down and fail. In most cases connector corrosion is the problem. Because the down leads carry power up to the device, the combination of current and moisture causes the failure. Connectors should be sealed with heat-shrink tubing and then taped.

7.652 Some systems today import or deliver programming from hub to node using a microwave link. There are basically two types of microwave systems. One is the familiar amplitude modulated link (AML) system and the second is the single-channel, up/down type.

The AML system operates in the frequency band of 12.2 to 13.2 GHz, known as CARS band frequencies. The up/down conversion is a short-haul system where the desired band is up-converted to a microwave channel at the transmitting end and down-converted back to the normal television band at the receiving end. One manufacturer provides a system that up-converts the whole cable television band to microwave and transmits it to a receiver that then converts the band back to its normal channels.

Once a microwave system is placed in normal operation, periodic checks on the transmitter and receiver should be made as a normal maintenance procedure. As klystron tubes or traveling wave tubes age, they deteriorate. Normal monitoring of the drive signals and power supply parameters will indicate the condition of the active elements. If a component is operating on the edge of normal electrical parameters, the component can be changed out before a breakdown in service results. Points to inspect are the down-lead cables for any pinching or crushing, antenna pointing brackets, grounding wires, and areas of general corrosion.

Typically a well-designed and installed microwave system will give many years of service, provided normal maintenance and replacement of components is performed. A common microwave facility outage occurs when the commercial power fails and the standby unit fails to work. This problem can be avoided with periodic cycle testing of the backup power system. Most other problems with a microwave radio link are due to weather damage such as flying objects, severe wind, rain, snow, and ice. Trees or construction of tall buildings blocking the line-of-sight path are two obstructions beyond control of the cable operator. The only cure is to relocate the transmit/receive sites to get a clear line-of-sight path. Replacing the microwave facility with a fiber-optic system is a possible solution.

7.653 Most cable systems have a number of satellite-receiving stations necessary to provide programming from a number of satellites. Breakdowns in any one of the systems can cause loss of many satellite channels. Getting the system back up and running is of utmost importance.

The most common problems are snow, ice, or debris falling into the dish antenna; corroded and damaged cable or connectors; and the effects of lightning. Snow and/or ice buildup in the parabolic dish antenna occurs in antennas with a large elevation angle. Some cable operators use a commercially available heating system that causes the ice and snow to melt and drain away. Debris such as trees, branches, and limbs can fall into the antenna and often damage the feed horn system. Systems that have problems with debris and trees need periodic inspections by competent technicians. Observations of the antenna system and the surrounding area are important. The satellite-receiving site should be visited at least twice per week, and the electronic equipment areas should be inspected as well. In general, a properly installed and maintained satellite system will give years of service.

Some systems still use pressurized, hollow coaxial feed lines, known as heliax. An air pump with a canister of a drying agent is used to pump dry air into the heliax. Breakdown of this pump occurs when the desiccant turns pink or when the signal from the antenna gets attenuated. When it gets too small, the satellite channel gets sparkles. This again is the result of poor maintenance.

Again, interference from other microwave facilities operating at the same frequencies can cause problems with signal degradation. Some sites may pass on site measurement at the initial construction time, but land clearing and/or nearby construction can eliminate the natural shielding effects. The cable operator then has two choices: relocate the site or construct an artificial shield. If the situation is not too severe, an antenna shroud or protective shield will effectively reduce side lobes of the antenna receive pattern to lessen the interfering signal level. Such situations are beyond control of the cable operator who must solve the problem.

Summary

Someone once said "to measure is to know" To measure the various performance parameters in a cable communications system, proper accurate instruments are required. Many instrument manufacturers have provided a plethora of instruments. The hand-held variety are most popular with field technicians. Much progress has taken place with hand-held instruments. The accuracy and the capability of battery operated hand-held instruments are available. Signal-level meters capable of measuring analog and digital data carriers are currently used by cable system operators. Portable spectrum analyzers with LCD screens are presently available.

Specialized cable test equipment that measures the distance to a cable fault (i.e., a cut, crushed, or damaged cable) is known as a time domain reflectometer (TDR). New, battery-operated portable TDR sets are available on the test equipment market. Also, cable locators that can accurately find the path taken by underground cable have proved very useful and time-saving. Using the cable location equipment with a TDR measurement, the location of a fault in buried cable can be determined accurately. This process minimizes the manual digging for the damaged cable.

As long as any coaxial cable is used in a cable communications system, signal leakage has to be minimized according to FCC regulations. This leakage-monitoring program has to be followed and the leakage levels and locations have to be recorded. This is required for the system to calculate its cumulative leakage index (CLI), which has to be less than the merit number of 64. Periodically the measurements have to be made and the CLI calculated in case an inspector from the FCC arrives and asks to examine the leakage records. Most systems repair leaks such as loose coaxial connectors as soon as they are found. Distribution equipment with the connectors and active and passive housing cause much of the larger and easier to locate leaks. Old drop cable that was used with early type "F"-connectors causes a lot of small leaks that can affect the CLI rating. Some cable system operators have a maintenance program that calls for systematically replacing old service drops.

Fiber-optical cable plant used as trunk cables require testing of the presence and level of the optical carriers. A hand-held optical power meter is the most cost-effective and commonly used instrument for cable operators' use in fiber-optical plant maintenance. These instruments can measure the optical signal level in dBm at the output of optical transmitters as well as at test points in the fiber-optical plant. In the case of a cut optical-fiber cable, an optical time domain reflectometer (OTDR) can be used to find the distance from one end of the fiber-optical cable to the break. Splicing of fiber-optical cable is time-consuming and expensive to do and requires a sealed closure system to protect the fiber splices.

Regardless of the cable distribution system, the base-band signal quality is an important parameter. Quality tests for analog signals are level, distortion, and signal-to-noise ratio. For digital levels the parameters are signal level, signal-to-noise ratio, and bit error rate (BER). Instruments able to measure these parameters placed at both ends of the transmission medium can measure the overall signal quality.

Most present-day cable systems use a standby system capable of automatically powering sections of the cable plant should a failure of commercial power occur. An alarm system indicating that one of the standby power supplies has gone into the emergency mode can send a signal back to the headend source, warning of the power outage. Regular testing and monitoring of the emergency power supplies is necessary, and it's also important to record the data. Such a program often requires a system technician to periodically inspect each standby power supply, usually at least twice per year. This inspection requires a visual inspection for rust and corrosion as well as a test of the charge status of each battery in the battery pack. Placing the supply in standby mode for a test and checking the battery charging system completes the test. A log of the test data can be used to identify possible upcoming maintenance problems.

Instruments and their proper use are necessary for good cable system performance. However, the quality of technician education is extremely important. Most cable system operators send their technical staff to participate in seminars and training sessions sponsored by the Society of Cable Telecommunication Engineers (SCTE). Technicians can become certified at various levels after successfully passing the required tests. Tests and measurements for most cable system operators involve testing on satellite systems, TV antennas, and/or microwave systems, as well as fiber-optical and coaxial cable distribution systems. Technician education in all of these areas is necessary.

Questions

1. Describe the basic operation of a signal-level meter.

2. Explain the basic principles of operation of a spectrum analyzer.

3. What is the purpose of cable and passive testing?

4. Describe what is meant by white noise and how it can be used as a test signal.

5. Describe the process of sweep testing a device.

6. If an off-air television broadcast station received at a cable system headend produces a noisy/snowy picture, what is the first measurement the cable technician should take?

7. What is the most important signal-level point for a cable television headend?

8. List some of the advantages for a cable system to use a fiber-optical trunk system.

9. List and discuss some of the problems with fiber-optical plant.

10. When a technician is required to measure the subscriber drop signal for both analog and digitally modulated signals, what kind of instrument should be used?

11. For a cascaded amplifier trunk system used in many cable systems, list the signal distortions that should be measured.

12. Explain the process of signal sweeping for a cable system.

13. When a signal outage occurs at a specified location by a subscriber, list the steps in the correct order a technician should follow to correct the problem.

14. Explain why a log of end levels can be useful as a maintenance tool.

15. Explain why system leakage is so important to a cable system and the need for FCC compliance.

Problems

1. For an amplifier trunk cascade, what should be the measured signal level of the television audio carrier if the video carrier level is measured and specified as +18 dBmV?

2. If a reel of cable with a specified length of 2,000 ft and according to the manufacturer has a loss of 1.6 dB/100 ft at 400 MHz, calculate the signal level at the reel end if +25 dBmV is injected at the opposite end.

3. If the output level of an amplifier is measured at the output test point as +8 dBmV and the test point is 20 dB less than the true output level, (a) what is the true output level? (b) if the amplifier gain is 20 dB, what should the reading be at the −20 dB input test point?

4. A 1.5-km length of optical cable is to be measured for loss. A light source of −1 dBm of optical power at 1,310 nm is injected at one end. If the optical power meter at the opposite end measures −2.2 dBm, what will be the loss in dB/km?

5. If a section of feeder cable becomes inoperative and it has been determined that there is a lack of system power, list the steps in tracking down the problem.

Cable System Network Design Considerations

Objectives

After learning the material covered in this chapter, the student will be able to

- Understand the development of the telephone system installed in this country.
- Describe the types of cables used in telephone systems.
- Explain the type of network topology used in the telephone system's outside plant.
- Explain the cable television network topology and the coaxial cable used in the outside plant.
- Describe the method of two-way operation used by many cable television systems.
- Explain the network topology of LAN/WAN commercial enterprise networks.
- Describe the use of Ethernet in LAN/WAN systems.
- Understand the basic principles of optical-fiber systems.
- Explain the application of fiber-optical technology in telephone, cable, and LAN/WAN systems.
- Describe the conversion process from optical signals to cable, telephone, and LAN/WAN systems.

8.1 Types of Cable Communication Networks

The two large cable communication systems in use today in the United States are the telephone system and the cable television system. These two systems today consist mostly of several large multiple operators.

The telephone industry did not start out in this manner; it consisted mainly of the American Telephone & Telegraph Company (AT&T). However, there were many small telephone systems that filled the needs of small, isolated rural areas. Even today there are still a few small systems in operation, but most of them sold out to AT&T over the years.

The telephone industry was formed to provide voice communications to its customers. AT&T, also known as the Bell system, was formed originally by the inventor Alexander Graham Bell. Being a basically frugal and moral man, his business plan contained many issues that promoted his venture. Quality of service (QOS) was an important issue to the fledgling Bell company. Bell allowed a certain percentage of his company's profits to be invested in research and development that would improve voice communication and the service quality. Bell Labs was formed to do the research. This research staff consisted of engineers and scientists recruited from the best universities. Bell Labs did notable work in researching human voice and hearing, which resulted in improved telephone equipment. However, published scientific papers on sound reproduction directed the way to high fidelity and stereo systems we enjoy today.

The work by Bell Labs has produced notable devices that have contributed to the commercial success of our society. Bell Labs' research in the field of semiconductors led to the invention of the transistor, which has been used for a great many commercial, military, and industrial circuits and systems.

Also of interest, the Bell system parent company determined that manufacturers weren't supplying the telephone company with equipment that could meet the required quality. Therefore, the Bell system people developed their own equipment manufacturer, the Western Electric Company. Western Electric made telephone sets, switchboards, and terminal equipment. Such equipment was designed to remain in service a minimum of 40 years.

The telephone company, or Bell system, has gone through difficult times that began with the Carter Phone court case, followed ultimately by the divestiture phase. However, the telephone network still provides voice communications as well as digital data communications to the commercial world and is a major carrier of the Internet.

8.11 The Telephone Network

The original telephone network consisted of cables mounted on wooden poles placed along the streets of a community. These poles were often called "telephone poles" even though electric system wires were mounted on the same poles. By common agreement the electrical wires used the space at the top of the pole and the lower regions were used for the telephone plant. About one foot above the telephone plant was space designated for use by the town that originally issued permission for the poles to be mounted along the streets in the first place. The towns and municipalities use this pole space to install the fire alarm wires. The telephone cables basically connected the customer service drop wires from the customer's home to a terminal on a pole where a connection to the telephone wires is made. The telephone cable wired the customer's telephone back to the community central telephone office. This office, presently known as the local exchange (LEX) office, was the termination for all the local system telephones. Because all calls had to pass through the local exchange office, the network topology was known as the switched star topology, which is shown in Figure 8-1.

For calls beyond the local exchange to another community, trunk lines connected the local exchange offices together. An elaborate method of trunk lines and various classes of offices, which acted as switching terminals, extended the topology of larger switched star networks connecting the many classes of offices. The LEXs were housed in a building often resembling a residential house. Inside there were switchboards manipulated by human telephone operators. With the development of pulse dialing, the switching function became automatic.

The long-distance trunk system first used coaxial cables that ran across the country, using single sideband radio technology. Many telephone channels could be carried on one coaxial cable. At various stations along the way the signals were amplified. Microwave radio methods were developed after World War II, and microwave wireless radio replaced the coaxial cable technology. The AT&T system made the decision in the 1940s to go digital, converting voice signals to digital data streams. The microwave systems could transmit the digital signals as easy as analog signals. For

Figure 8-1
The control center

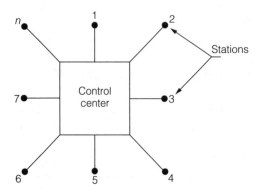

The control center acts as the switch allowing communications between stations 1, . . . , n. The control center for a commercial LAN would be the smartrouter. For a telephone network, the control center would be the local exchange.

transcontinental and overseas service, early submarine cable was abandoned in favor of satellite-relaying techniques.

8.111 Basic twisted pairs of copper wire are still used today to connect subscribers' telephones to the system. This is often referred to symbolically as the last mile. The twisted-pair cable made back in the early 1900s consisted of a pair of twisted copper wires insulated with silk threads impregnated with wax. The silk threads were color coded so when a larger number of pairs were again twisted into a multipair cable, each pair could be identified. The whole multipair cable was wrapped with wax paper. The outer cover was made of metallic lead. This type of cable was made before the invention of the plastics industry. Such cables appeared as 25-, 50-, and 100-pair cables. Lead-covered cables were very heavy and were hung on galvanized steel messenger strands running from pole to pole. Wire loop rings were used to hang the lead cable to the strand. Splicing and repairing lead-covered cables required the expertise of working with molten lead and solder, similar to what early plumbers had to do with lead drain systems. The telephone company performed all its own maintenance, installation, and repair. The Bell companies treated their maintenance and installation personnel with good pay, benefits, and respect.

Lengths of twisted-pair cable had a characteristic impedance around 600 ohms, which was the standard audio impedance used in radio station equipment. Twisting the cable reduced noise and cross-talk between pairs. However, the distributed capacity caused the pair to act like a low-pass filter. To flatten the frequency response of the line, inductances were added every so often along the way. The value was determined by the value of the accumulated line capacitance and color coded to facilitate easy installation.

8.112 As stated earlier, the chosen network topology is called the switch star topology and is a natural selection for the telephone application. Each local exchange switching office is at the center of the star. Trunk lines connect the local exchanges to primary switching centers and then connect to toll-center switching centers. This type of topology is shown in Figure 8-2. The star topology supports the concept of switched circuits that set up the placing of a telephone call between two telephones. These circuits remain connected until the call is completed by one of the parties hanging up the telephone. Since the telephone company converted to digital technology, all the switching algorithms that set up a telephone call are done using computerized switching under software control. This method has the added benefit of keeping track of circuit routing, time, and

Figure 8-2

The telephone-switching center network

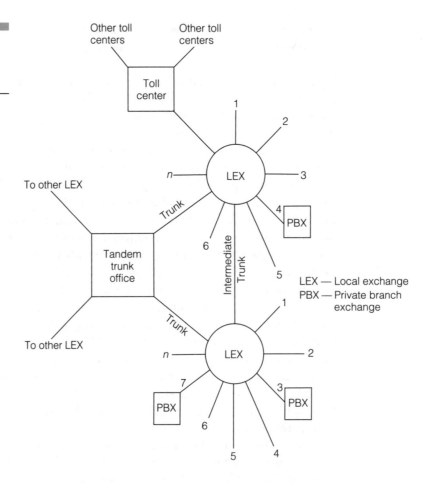

charges—all used in the process of billing the customer who made the call. The circuit switching and signaling system is known as the ESS5, the 5 being one of the latest.

8.113 The telephone network in this country has always provided improved quality and service. Small independent companies sold out to the larger ones, forming huge consolidated telephone systems. Still, most of the lines along the public ways, both aerial and underground, belong to AT&T or their former affiliated companies, known as the regional Bell operating companies (RBOCs). Sprint and MCI formed companies that offered similar services but didn't own or control much of the physical plant. AT&T invested in the research and development of fiber optical technology. The system, known as SONET (synchronous optical network), was developed. AT&T basically converted its trunking system to optical fiber. Digital switching and digital data words (bytes) were transported in packets in the SONET systems payload. For students further interested in SONET, Appendix F should be of interest.

Optical-fiber technology provides a higher signal frequency bandwidth, which allows high-speed data and voice communication over long distances. Optical submarine cable installed across the Atlantic and Pacific oceans has lowered the costs of intercontinental voice and data transmissions. AT&T was one of the major investors in many under-sea fiber optical projects. Single-mode optical cable using laser transmission technology and sensitive photo PIN diode receivers are the basic long-distance communication devices. The Internet uses high-speed data communications, which makes it possible for everyone's computer to be connected to all types of servers, giving users access to a multitude of information and products for sale, as well as movies, games, and audio programming. Still, it is basically the telephone company's high-speed data infrastructure that is running the Internet.

8.12 Cable Television Systems

As discussed in Chapter 1, cable television systems used coaxial cable as the transmission medium. Coaxial cable, with its wideband width, could accommodate the whole VHF television band, with the stations on their assigned broadcast channels. Early systems often frequency shifted the high VHF channels 7–13 to the low-band 2–6 because the cable's attenuation was less. Improved coaxial cable was quickly developed when the manufacturers realized this large market. The improved cable allowed the high-VHF channels to remain where they were originally. The early cable systems began as five-channel systems and single-cable delivery from the transmitting headend to the subscriber's home. Improved cable and electronic amplifiers made possible 12-, 21-, 30-, 35-, 52-, 60-, 87-, and 102-channel systems. Some cable systems offered improved FM radio service and used the broadcast 88–108 MHz band intact.

8.121 Coaxial cables originally were small sized. The feeder cable was 0.412 inch in diameter and the trunk cable was one-half inch (0.500). Both types had a solid aluminum sheath (tubing) and a solid-copper center conductor. Often no jacketing was necessary in many areas of the country. Jacketing of PVC was used for coastal environments to protect the aluminum sheath from corrosive salt air. Underground, or direct burial, cable was developed by adding a flooding compound under the PVC jacketing. As a money-saving option, many manufacturers offered cable with a copper-clad aluminum center conductor, which had an increased loop resistance, to be used where the cable powering design did not need a low loop resistance.

Typically, the drop cable supplying service to subscribers was 75-ohm RG-59 flexible coaxial cable. The flexibility was obtained by using a braided copper shield. The center conductor was often steel, which gave added strength, with a copper cladding. Because drop wire, or cable, as it was known, was always jacketed, it was necessary to protect the braided outer conductor or shield. As time passed and the cable television industry matured, it was found that the drop wire had inadequate shielding against signal leakage and noise interference caused by ingress. Manufacturers of drop cable introduced a layer of aluminum foil placed under the braid, which was very effective in improving the shielding effectiveness. Further improvements added more layers of aluminum foil and another layer of braid, which vastly improved shielding as well as mechanical strength. By present-day standards, the drop cable used is RG-6 type, which is double shielded, has a larger diameter than RG-59, and has lower attenuation and greater mechanical strength. As drop cable improved, so did the connectors. At first, the simple type "F" connector used by the master antenna industry (MATV) was adopted. This was a very crude and poor connector and caused a great number of service calls. Because this was very costly to cable operators, new types of "F" connectors, with improved crimping sleeves, were developed. New drop cable stripping tools and hand-crimping tools made the installation of the new "F" connectors easy and quick, which produced a vastly improved connector at manageable costs.

Improvements in the manufacturing process and the cable operators' need for lower loss cable enticed more cable manufacturers to enter the market. Feeder cable size increased from 0.412 inch, commonly known as 412, to 0.500 inch, or simply 500. The dielectric material improved, and cables appeared on the market with gas-injected polyethylene foam between the aluminum sheath and the center conductor. Also, one manufacturer used a polyethylene disc where the center conductor is held concentric with the aluminum sheath. Dry air acts as the dielectric insulating material. This type of cable exhibits less loss as the operating frequencies increase. Many cable systems use 0.500 inch as feeder cables and 0.750 inch (750) or 1.0 inch (1,000) as trunk cable. Cables are also available in 0.65-inch and 0.875-inch sizes, 0.65 for feeder systems and 0.875 for trunk systems.

8.122 The topology of traditional cable television systems was determined by the developing industry as needed to deliver as much television programming as possible. Because the signals were gathered and assembled, hopefully at one location (the headend), to be delivered to subscribers distributed throughout the service area, the topology was called a tree network. The trunk system transported the signal to the ends of the service area. Along the trunk, bridging amplifiers allowed the tree to sprout branches (feeders) that contained the taps with the subscribers' tap port. Essentially such systems started out as a delivery method of providing television programming to subscribers. Little thought was given to forming a reverse signal path back to the headend. Some cable operators considered providing meter-reading services for water suppliers or electric companies, and some operators envisioned offering security or alarm services. Such a return path now inverts the tree to branch topology as the reverse tree topology. Consider the fact that noise generated at the headend increases down the trunk according to the accumulation of noise specified by each amplifier's C/N specification. At a bridger point the feeder adds more noise contributed by each distribution amplifier, so at the subscriber's house, the C/N is at acceptable limits necessary to provide good clear pictures. Now going in the reverse direction from all subscriber locations back to the headend through a cascade of reverse amplifiers, all the branches supply noise that is combined in the reverse trunk. Where the trunk splits going downstream, it combines going upstream so at the headend the C/N is often unacceptable. This can be calculated approximately but it can be easily measured using a spectrum analyzer looking at the upstream frequency band. Some system operators limit the number of locations using a channel in the reverse band. That strategy limits the noise ingress in locations not requiring upstream services.

Some progressive systems found that what was needed was more reverse channels and fewer downstream channels. The mid-split method allowed nearly half the channel upstream and a bit more than half in the downstream direction. Not many cable operators found this attractive financially, because downstream services are in more demand by consumers.

8.123 Remember that so far, cable television systems operated as cascades of broadband amplifiers in cascade both upstream and downstream. These cascades of amplifiers were power hungry and required periodic maintenance. To overcome such costs and keep the cable operators in business, the upstream and downstream channel capacity had to be used to produce revenue.

Many cable television operators envisioned their plant could contain a multitude of services to and from subscriber terminals, thus increasing more revenue. Internet service was one application, and with a cable modem with an operating program supplied, more rapid communications resulted. The bit rate was usually about 1.5 Mbps (T-1 rate). This could diminish to nearly 250 Kbps, depending on the number of participating subscribers the operator served. Naturally, VOIP may be the next stop.

8.124 What started out as a tree-branch cable system, through changes in network topology and equipment, evolved into a combination of ring-bus bidirectional hybrid fiber-coax (HFC) system supplying a variety of services to the subscriber. The coaxial plant operates as a spectrum of RF-modulated carriers, which is classified as frequency division multiplexing (FDM). This whole spectrum is converted to light energy and carried on optical fiber. Some systems placed a group of carriers on one fiber and another group on another fiber. Also, it was found that one group of carriers could be converted to light energy on an optical carrier and carried on a fiber, and other groups of RF carriers could be converted to another optical carrier, and all could be transmitted on one fiber. This process was known as wavelength division multiplexing (WDM). This method, shown in Figure 8-3, was efficient in the use of an optical fiber.

How a cable television system evolved into an HFC network depended on how the coaxial cable RF plant started out. Typically the off-air antenna location was the source of the company's product—namely, television programming. With the advent of satellite television technology it was often prudent to place the satellite-receiving antennas at the tower site. In some cases, this was not possible due to permit problems, access, or interference. Consequently a headend node had to be formed, which collected all the television signals from the off-air tower and the satellite-receiving station. This node was often placed at the site of the company office or the local television studio site. The main subscriber trunk cables were fed from this headend node. Some of these topologies are shown in Figure 8-4. The best case is where all the signals originate at one location. Few cable operators are this fortunate. The next best situation is where most signals originate at another, remote location. The remote site, in many cases, is connected to the main signal source node by a short cascade (two or three) of RF amplifiers through a separate trunk cable. Cable operators would replace this coaxial cable amplifier cascade with an optical-fiber system, which took less power and required less maintenance.

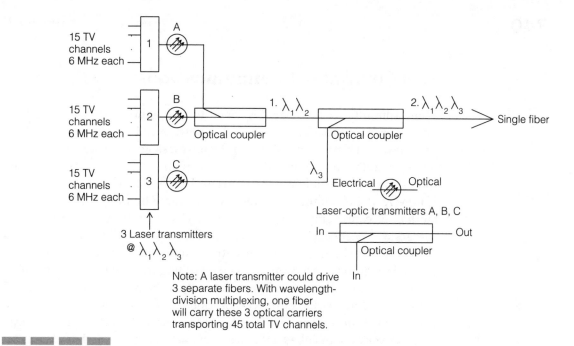

15 TV channels 6 MHz each

15 TV channels 6 MHz each

15 TV channels 6 MHz each

3 Laser transmitters @ $\lambda_1 \lambda_2 \lambda_3$

Optical coupler

1. $\lambda_1 \lambda_2$

Optical coupler

2. $\lambda_1 \lambda_2 \lambda_3$ → Single fiber

λ_3

Electrical — Optical

Laser-optic transmitters A, B, C

In — Out

Optical coupler

In

Note: A laser transmitter could drive 3 separate fibers. With wavelength-division multiplexing, one fiber will carry these 3 optical carriers transporting 45 total TV channels.

Figure 8-3
Example of wavelength-division multiplexing

Figure 8-4
Network topologies

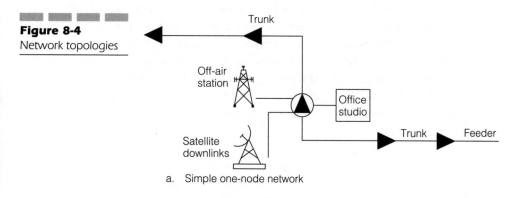

Trunk

Off-air station

Satellite downlinks

Office studio

Trunk   Feeder

a.   Simple one-node network

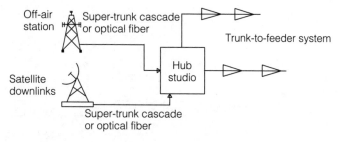

Off-air station   Super-trunk cascade or optical fiber

Trunk-to-feeder system

Satellite downlinks

Hub studio

Super-trunk cascade or optical fiber

b.

The hub is the main signal distribution point, fed by remote broadcast station receivers and satellite downlinks.

8.13 Computer Communications

When computer technology developed to the point where personal computers became practical and affordable, the need to interconnect them became a necessity, particularly in the business world. Naturally the telephone system was investigated, and telephone modems were developed that made possible data communications between computer workstations. However, within commercial offices in the same building a more efficient method of connecting workstation to workstation to mass data storage servers was needed. Companies that had their own PBX telephone system could use telephone modems for data communications. Many companies decided to interconnect computers by installing their own networks, called local area networks (LAN). These networks were designed as hard-wired bidirectional networks. Large corporations with many offices spread over several cities developed methods to interconnect the LANs. These large data networks were called wide-area networks (WAN). Further expansion gave rise to metropolitan wide-area networks (MWANS).

8.131 The network topologies for these networks were basically one of three types: the star, ring, or bus configuration. Each had its own access method. Figure 8-5 illustrates the basic network concepts. The public-switched telephone network (PSTN) is basically a star network with the local exchange office (LEX) at the center of the star where circuits were switched. IBM proposed the ring network, where a token was circulated around the ring. The station that had the token had permission to transmit to another station. This type of system is known as a token-ring net-

Figure 8-5
LAN topologies

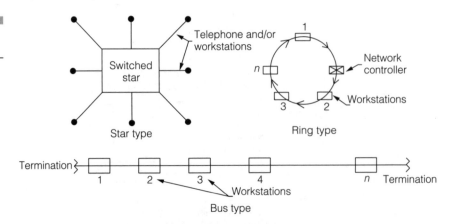

work. This token was actually a common password set by the network operating system software. The bus topology uses a simple cable system of either UTP or coaxial cable connecting all workstations. The ends of the cable have to be properly terminated in the characteristic impedance. Access is on a first come, first served basis and the procedure uses carrier sense multiple access/collision avoidance (CSMA/CA). The station wishing to transmit to another station on the bus listens to the bus to check if anyone is transmitting. If not, the bus is available; if a carrier is sensed the station waits and tries another time. If a rare occurrence of two stations transmitting simultaneously happens, both stations detect each other's carrier and go into a time-out mode. The time-out period is random for each station and when it is over they can try again. The station that has the shortest time-out period gets to try transmitting before the other station. Data can be transmitted between stations at adequate rates and periods until the number of users (workstations) increases to a point that data-handling speeds decrease.

8.132 Ethernet local area networks are one of the oldest and robust bus systems still in use today. Many say that Ethernet keeps reinventing itself. Data rates kept rising with each new issue of Ethernet. Xerox and Digital Equipment Corporation, along with Intel, collaborated in forming and promoting Ethernet. The Institute of Electrical Electronic Engineers (IEEE) worked with the founding corporations to sponsor the 802 standards group. From time to time, as faster semiconductor digital circuits were developed, the 802 standards would be upgraded and Ethernet data rates increased accordingly. Ethernet started to speed up at 10 based T, which was at a 10 mbps. Coaxial cable at a 50-ohm characteristic impedance (RG8u or RG58u was used in a basic bus topology). Ethernet topology was introduced in Chapter 5 along with other LAN topologies. The principle reason Ethernet has lasted so long and is presently supporting data rates of 10 Gbps is that it is a protocol where the data frames of packets are defined. As cables improved the simple frame format would easily support the high data rates.

8.133 LAN/WAN cables have improved over the years. Unshielded twisted-pair cables used by the telephone industry were the first basic carrier of digital signals. When one pair would not support the speed requirements another pair was used, resulting in CAT 3, 4, 5, and 6 series of paired cable. Coaxial cable, with its larger bandwidth and hence high speed, allowed for higher data rates. The telephone industry supplies services to the subscribers using the twisted-pair copper cables. The cable television industry uses 75-ohm coaxial cables to supply its services to subscribers. Using coupling pairs of CAT-5E cable using DSL or ADSL technology the telephone system offers the subscriber data rates of 1.5 Mbps. The cable television systems are supplying similar data rates, along with many television program channels. Both the telephone systems and cable television systems are using optical fiber as their backbone system, and the final mile serving the subscriber is the weakest link.

8.2 Optical-Fiber Applications

Optical-fiber applications have appeared in corporate and municipal LAN/WANs, telephone networks, and cable television systems. Whenever a communication network requires more information thru-put, optical-fiber technology in most cases is employed. As stated in Chapter 4, optical fiber is not as easily connected as metallic type cable. Splicing of optical fiber is extremely important for low-loss optical cable systems. The process of fusion-splicing for low-loss splices is painstaking. However, nearly automatic fusion-splice equipment is available, although quite expensive. Splitting and coupling to optical fibers involves splicing together the optical fiber to the splitting devices usually installed in splicing trays in cable enclosures.

8.21 Telephone Optical System

The telephone network (namely, the Bell system) was the first to use optical fiber. After all, a lot of the research on optical fiber was performed at Bell Labs. Also, the synchronous optical network (SONET) was developed at Bell Labs. Essentially SONET describes the digital hierarchy as well as the packets and frames of the transport system. The SONET-type transport system could use any high-speed non-optical cable system if the cable or radio system had appropriate bandwidth. However, because single-mode optical fiber supports gigabit rates it is the transport method of choice. The telephone system first utilized optical fiber in the telephone lines connecting the various switching systems located across country. Digital signaling methods were developed at Bell Labs from the laws developed by Shannon and Hartley. The result of Bell Core early research on digital signals was the decision for the telephone system to go completely digitally. Because the main product of the telephone company was voice communications, the speech waveforms were converted to a string of digital words. The T-1 carrier, consisting of 24 voice channels, combined into a 1.544-Mbps data stream. This digital methodology was used with the twisted-pair coaxial systems as well as microwave links used by the telephone network. With the development of optical fiber and SONET, the whole telephone trunking system was replaced. In later years, the telephone system has added optical fibers to the local exchanges and in some instances to corporate office campuses. Even today the telephone handset is connected to the telephone network by metallic wires.

The telephone network is the largest carrier of digital data, mainly the Internet. After all, if a voice circuit carries a series of 8-bit bytes and a computer uses a 16-bit word (28 bit bytes), both of these signals can be carried on any network transporting 8-bit bytes methodology. The SONET transport system lends itself very well to carrying voice, video, and data signals. The telephone system's vast SONET system carries the Internet to the end users. The SONET system is operated by parts of the old AT&T Bell network, known as the regional Bell operating companies (RBOC) or Baby Bells.

To institute the supervision and control of data transported over the telephone system, the integrated services digital network (ISDN) was developed. ISDN is an elaborate and complicated protocol that allows digital data to be transported through the telephone network, which is mostly optical fiber. Higher orders of digital signaling rates, DS1–DS3, are used. The international standard organization CCITT has adopted the ISDN concept.

8.211 The telephone system topology is basically a switched star–type network; the star networks were connected together by linking the star centers together to form larger stars. This type of network connection was called trunk lines and/or long-lines. The first application of fiber-optical topology was to replace the wire type trunk lines with optical fiber. This method is illustrated in Figure 8-6. Optical fiber use was extended deeper into the telephone networks to connect all the local exchange offices. These offices presently have decreased in size principally due to the use of integrated circuit–type switches. Therefore, many central offices had green steel cabinets placed along the public streets. Connections to local homes, offices, and commercial businesses are connected to these unmanned automated cabinets where the local control of the telephone system exists. The fiber-optical trunking system operates using single-mode optical fiber. At the higher class switching centers these digital telephone signals are further multiplexed into SONET.

Figure 8-6

An optical-fiber
network replaces
trunk lines

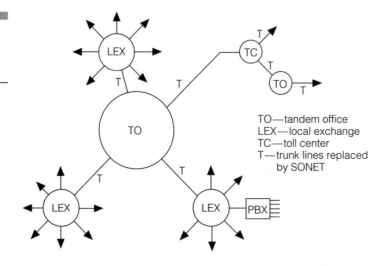

TO—tandem office
LEX—local exchange
TC—toll center
T—trunk lines replaced
by SONET

8.212 Optical cable technology connects the telephone communication system to the local exchange office, referred to as a local central office. Here, the local telephone services are converted to digital signals at the DS-1 rate of 1.544 Mbps. These signals are further multiplexed to DS-2, DS-3 rates according to the high-level data link control (HDLC) standard developed and used by the telephone industry. The HDLC standard is the communication protocol and is accepted internationally as the protocol for digital communications. The X.25 packet switching protocol developed by the International Telecommunication Union (ITU) uses the HDLC protocol in carrying digital data over public data networks. The HDLC protocol dictates that standardized digital frames will packetize the digital data or telephone signals. A basic HDLC frame format is shown in Figure 8-7. Because the voice signals generated by the subscriber's telephones are converted to 8-bit bytes, they look like any 8-bit data bytes. Therefore, header, trailer, and error control bytes have to be used so these digital signals can be properly reassembled and converted to analog (voice) signals at the central office serving the received party. HDLC protocol is

Figure 8-7

HDLC frame format
adopted by telcos

Start flag	Address	Control	Information	FCS	End flag
8 Bits 1 Byte	8 Bits 1 Byte	1 Byte	Variable	2 Bytes	1 Byte

complicated and exacting in its task of transporting data over the public switched telephone network (PSTN).

8.213 The development of SONET was needed to transport high HDLC data rates over optical fiber. Also, the integrated services digital network, developed by AT&T, was the method used for transporting digital data and digital telco systems through the PSTN system. To speed up the digital data packets, SONET was invented. The SONET system has been standardized by the American National Standards Institute (ANSI) and adopted into the synchronous digital hierarchy of the ITU.

Because SONET is a high-speed system, slow data has to be multiplexed up before entering the SONET domain. DS-3 is the entering digital data rate accepted by SONET systems. DS-3 operates at 44.736 Mbps, which can fit into the SONET optical carrier 3 (OC 3) data rate of 51.84 Mbps. SONET systems are still in use today, although the equipment has been significantly upgraded and improved.

8.22 Cable Television System Applications

Cable television operating systems took advantage of fiber-optical technology to basically replace the trunk-cable part of the distribution system. This application essentially identified distribution areas where subscribers were located and formed nodes where the optical signals were converted to the RF television spectrum. The RF coaxial plant served subscribers from the usual tap devices.

8.221 Many cable systems either overlashed or exchanged the coaxial plant with optical fiber to transport the cable signals to the subscriber distribution points. In the case of underground cable plant, optical fiber was installed either in existing ducts or conduits or newly installed. Although expensive, the decision to replace the coaxial trunk cable with optical fiber paid off in several ways. Most importantly it decreased the number of amplifiers in cascade, thus reducing signal degradation, system leakage, and electric power consumption. The signal improvement resulted in lower noise, less distortion, and more resistance to the effects of temperature. Many systems reported positive comments from subscribers. Many cable operators followed this strategy of replacing the trunk system with optical fiber, so new instruments were required to test and maintain the new optical plant. Also, the technical people had to learn how to test, repair, and perform maintenance procedures on the new fiber-optical terminal equipment.

8.222 When the coaxial trunk cable was replaced by optical fiber, the reverse or upstream signals had to be addressed. Systems that were basically one-way had no problem. For systems that had active two-way communication, there was a choice of maintaining and keeping the old coaxial trunk and going to a low frequency band just for the reverse system. Because this choice offset the respective power cost savings and required continual leakage testing, it was better to use the optical system for the reverse as well. Using optical fibers in the optical cable was a choice for cable operators. Therefore, for each node, two separate fibers were needed, one for forward service and one for reverse. This choice was available to the cable operator because optical cable contained several fibers in a given cable. Many cable operators installed cables containing many more fibers than was needed. Several industry experts have said that there is an abundance of installed fibers by many companies throughout the country.

The dense wavelength division multiplexing (DWDM) technique uses many optical carriers of different wavelengths (frequencies) on a single fiber. The same fiber could support both the upstream and downstream RF frequency spectrums; therefore, one fiber can provide the channel capacity for a cable operator. Figure 8-8 illustrates the applications mentioned.

8.23 Cable Node Placements

Cable operators had to select the location where to convert from the optical domain to the RF domain and vice versa. These locations were often referred to as nodes and were selected according to subscriber density. The upstream system often used the common sub-split reverse band of 5–30 MHz. Many of the large MSOs' reverse systems used digitally modulated carriers (QAM), so the 5–30 MHz band was adequate for the

Figure 8-8

Optical signal combiner

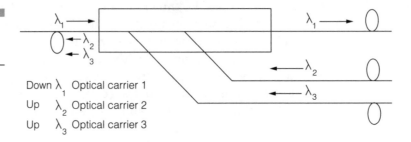

Down λ_1 Optical carrier 1

Up λ_2 Optical carrier 2

Up λ_3 Optical carrier 3

needs for reverse communications. This node development is graphically described in Figure 8-9.

8.224 The nodes where the cable system converts from optical cable to coaxial cable may be aerial plant or underground plant. Manufacturers of such equipment make the equipment available in both strand-mounted enclosures or pedestal- or vault-mounted enclosures. This equipment basically consists of an optical PIN diode detector connected to a broadband RF amplifier. From this amplifier the whole forward frequency spectrum can be distributed to subscribers by the coaxial cable distribution system. This concept is shown in Figure 8-10. Power for the optical-fiber system to the RF converter is obtained from the normal coaxial cable plant at 60 or 90 v.a.c. Remember that optical cable does not carry low-frequency a.c.—or d.c. for that matter. However, some optical cable contains a metallic conductor that could be used for powering.

For the reverse system, the subscriber's upstream information is multiplexed with other subscribers' data onto a sub-split reverse carrier. At

Figure 8-9
Upstream/downstream signal combined in a single fiber

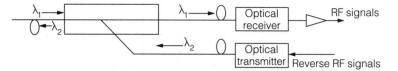

Figure 8-10
Optical RF node, with optical fibers A and B and RF terminals 1 and 2

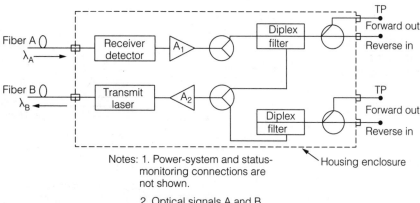

Notes: 1. Power-system and status-monitoring connections are not shown.

2. Optical signals A and B could be combined by an optical coupler to one fiber.

each node the upstream RF channels modulate an optical transmitter of a different wavelength from the downstream wavelength and are coupled to the optical fiber by an optical coupler. A simplified diagram of this technique is shown in Figure 8-11. Because the choices for many variations of this technology are many, the most common are illustrated. In most cable television system applications the headend signal source contains the downstream system optical transmitters and the upstream optical receivers/demultiplexers. The nodes' optical receivers convert the downstream signals to RF and provide optical transmitters on the upstream path. In between there is no device of any kind unless it is an optical power divider. This type of system is sometimes termed a passive optical network (PON).

The variations in optical-system design used by cable systems employ several methods. At the primary headend, if the first optical cable length is long, the choice would be to use several RF-to-optical transmitters operated at different wavelengths driven by 18 to 20 RF channels. This approach allows a higher drive signal level to the laser transmitter, resulting in a larger received signal at the end of the run. Another approach would be to use a high-output transmitter and carry a greater number of channels. A change in the network topology in many existing systems is often out of the question due to increased costs and downtime. Most systems have been in existence many years and have undergone several rebuilds where optical fiber has been added as the main signal transportation method. Manufacturers of optical cable and electro-optical transmitters/receivers often have a customer applications department offering cable operators with design assistance. Such services are often without cost. Most manufacturers of cable system optical equipment carry a good

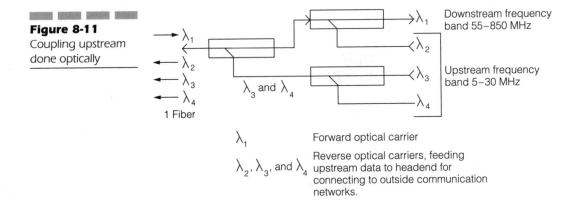

Figure 8-11
Coupling upstream
done optically

amount of devices to fit a customer's need. Such an approach is shown in Figure 8-12.

8.23 Local Area Connections

LAN/WANs have gone through several generations of cables. Older systems started using telephone-type twisted-pair cables operating in Ethernet. When the network expanded and higher data rates were required, conversion to a coaxial cable system resulted. LAN/WAN systems were not as long as either the cable television or telephone systems. The bus-type coaxial cable operating in Ethernet lasted a long time. However, many LAN/WAN systems required higher data rates, so optical fiber was selected to replace coaxial cable systems. Multimode optical fiber supplies the needs of many LAN/WAN systems. Today, a good number of systems

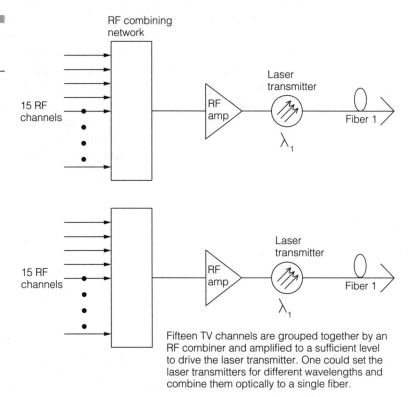

Figure 8-12
Combining multiple channels

Fifteen TV channels are grouped together by an RF combiner and amplified to a sufficient level to drive the laser transmitter. One could set the laser transmitters for different wavelengths and combine them optically to a single fiber.

operate with 62.5/125 or 50/100 multimode fiber. Because this type has a distance speed limit, single-mode fiber bridges provide the linking of several multimode systems.

8.231 LAN/WAN systems that convert to optical fiber systems often have the fiber installed in the same raceways and/or conduits as the coaxial cable. When the new fiber system was completed and workstations and servers were connected, they would end the upgrade process and the old LAN cable was left in place. Present-day building codes now require the old, unused cable be pulled out. This requirement was the result of the new fire protection rules for buildings. Many state laws require a company to remove all of its cable system if they move or vacate the premises.

8.232 A great many LAN systems operate using Ethernet and when optical fiber is used, Ethernet can be upgraded from 100 base T to 10-G bit Ethernet. The whole LAN operation is usually speeded up by a factor of at least 10 over wired systems. In many instances more workstations are added as the volume of the user's business increases, with no problems with the LAN system operation. Each workstation connected to the optical system requires a special interface unit where the optical signal is converted to serial/parallel electrical signals required by the computer. Electrical data signals have to be converted to optical signals in order for each workstation to transmit. This procedure is shown in Figure 8-13. Several companies manufacture and sell such network interface units (NIU) to users and often provide design services as part of a package. Fast Ethernet systems send data packets containing digital information fast enough to provide video and voice over their networks. Commercial busi-

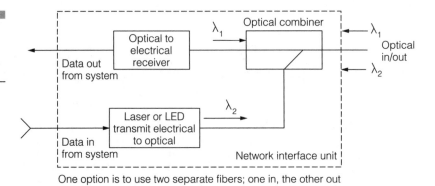

Figure 8-13
LAN to workstation network interface unit

One option is to use two separate fibers; one in, the other out

nesses using such Ethernet optical-fiber systems can provide audio/video channels for personnel training.

8.233 A ring-type network topology offers some advantages over bus LANs. The ring system uses the cable in a ring pattern, as shown in Chapter 4. However, remember that the ring system requires a token passing for permission to transmit through the system. A double ring system consisting of a dual-fiber cable in a ring pattern can carry high-speed data carrying an enormous amount of data. The fiber distributed data interface (FDDI) operating at 100 Mbps standard was adopted by ANSI and provides high-speed data up to 2 km. Today, that distance could be increased because many improvements in optical equipment have taken place. This ring system uses two fibers, one operating in one direction and the other in the opposite direction; this is often called a counter-rotating system. Such a system can be nearly self-healing in the case of mishaps. This FDDI dual ring is shown in Figure 8-14.

Figure 8-14
Fiber-distributed data interface

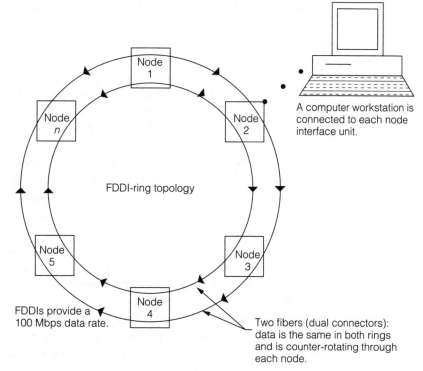

A computer workstation is connected to each node interface unit.

FDDI-ring topology

FDDIs provide a 100 Mbps data rate.

Two fibers (dual connectors): data is the same in both rings and is counter-rotating through each node.

Commercial businesses in most instances have their LAN systems connected to the Internet through either the telephone system (DSL/ADSL modem) or the cable television system (cable modem). Only high-speed LANs can take advantage of high-speed Internet connections. Slower LANs, of course, can connect through the ordinary dial-up modem.

8.3 The Interconnect Process

Cable television systems that have upgraded to fiber optical technology have improved their position as a communications provider. This technology is referred to as hybrid fiber coaxial (HFC) systems, and it provides television as well as high-speed data and voice services (VOIP) to commercial and residential users. Optical-fiber technology is the backbone of commercial LAN/WAN systems; as it connects to nodes where CAT 5, 6, or coaxial cable provide service to user workstations.

8.31 Cable Systems Interconnect

Many existing cable communication systems built today need to be interconnected by some means. This interconnect process will improve communications between systems, making possible HFC multiservice WANs. Cross-connecting of these many systems presents the problem of translation between network operating protocols. Most commercial enterprise LAN/WAN operate under Ethernet, while the telephone-based systems use frame relay and asynchronous transfer mode (ATM) protocols along with the synchronous optical network (SONET) technology. Digital information appears in 8-bit (byte) increments, which makes feasible interconnections via packetizing and framing translation between networks. Remember that telco digital communications follow the digital signal (DS) standard bit rates and are transmitted as bit streams consisting of frames of bits. These bit frames can easily form ATM data cells as well as SONET data cells. All of these methods utilize a two-way, forward/reverse, or upstream/downstream service for transporting data. To better adapt all the telco protocols for frame relay and ATM, some high-speed interconnect using single-mode optical fiber technology was apparent. Therefore, SONET was developed and put into service. The SONET cells were developed to accommodate the telco digital protocols. Digital signals carrying voice, data, and video were carried in SONET cells. As the saying goes, bit are bits, and as long as the bytes are accounted for in the SONET, all goes well. But as time goes by, everything gets old, SONET included. Next-generation SONET is on the drawing board and rest assured will be completely compatible with old SONET. The interconnection of systems is illustrated in Figure 8-15.

8.311 When a coaxial-based system is replacing its coaxial trunk with optical fiber, planning the location and network position of the optical-to-RF and RF-to-optical nodes is important. The positioning of such nodes in the system can drastically affect the cost of the upgrade. The nodes are often referred to as the upstream/downstream converters. Several manufacturers of electronic equipment designed for such nodes often provide little or no-cost services to the prospective buyer. Cable system operators must remember that all of the paying subscribers are connected to these nodes. The procedure often starts with a system map, where the location

Figure 8-15
A SONET loop

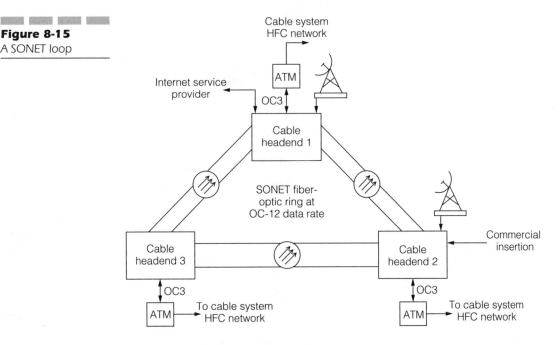

SONET fiber-optic loop connects headends to Satellite earth stations, the internet (VOIP), and commercial advertisement inserts.

of the distribution nodes can be placed. Some systems have skeleton overlay maps, eliminating a lot of detail. This type of map lends itself to placing conversion nodes. This concept is shown in Figure 8-16. Use of this system map is often referred to as a tree diagram. The number of fibers in the optical cable is an important factor. But remember that the largest cost is for the installation and placement. One fiber can be used for forward at one optical wavelength and the reverse at a different optical wavelength. The other choice would be to use a separate fiber for the upstream and another for the downstream.

8.312 Adding optical fiber as a cable system backbone makes some interesting applications possible. System performance can be monitored, and some signal switching can be accommodated by the use of one of the available fibers. Present systems can use an optical fiber as the status-monitoring reporting line. A system using such a monitoring and con-

Figure 8-16

Fiber-optic distribution system connects RF coaxial cable nodes to subscribers

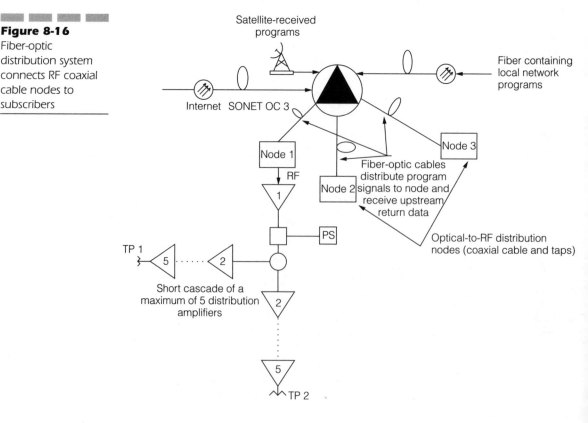

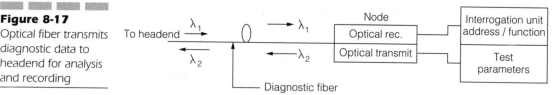

Figure 8-17
Optical fiber transmits
diagnostic data to
headend for analysis
and recording

Interrogation can be "on command" or "periodic."
Test parameters are:

1. Power supply output measurement
   of voltage and current.

2. Optical signal level of λ_1 and λ_2
   RF signal level (up/downstream)
   data is analyzed at the headend.

trol channel can make their cable system nearly self-healing. If not, the information received can direct the maintenance and repair staff to the problem area, thus minimizing downtime. Utilization of a diagnostic fiber is shown in Figure 8-17. This fiber contains the diagnostic results and transmits the data to the headend or company office for analysis. The diagnostic data analysis and compilation using a computer program provides management information on system problems and quality of service to the subscribers. Many cable operators have rebuilt their systems to provide voice, video, and data services. Such services consist of digital television (DTV, ADTV, and HDTV), voice over the Internet (VOIP), and Internet data and shopping services. At present hybrid fiber coax (HFC) technology is used here in America. However, Japan is extending optical fiber to the premises, making the entire distribution system optical. This means the system bandwidth is enormous, right to the user. When this will occur here in the United States, no one exactly knows. All eyes are on Japan.

Summary

There are three communication networks in use today that started with different purposes. The telephone system, which is the largest communication network, is the oldest and was for the purpose of voice communications. This network was expanded throughout the country so there were very few people without telephone service. With the development of television broadcasting stations, people desiring service would use a receiving antenna connected to their TV set. In difficult areas where the signal transmitted became weak, people received poor television reception. Rudimentary cable systems were able to install an antenna system to receive a good, strong stable signal and deliver it to subscribers by a coaxial cable network. This network was further extended and developed into a broadband cable distribution system providing many more kinds of services, not just television. With the development of two-way technology, many cable systems became a provider of voice, video, and computer services. With the addition of system-status monitoring and computer monitor programs, such cable systems became almost self-healing. Now with voice over the Internet, such systems were able to supply what the telephone system offered. The telephone industry responded by offering high-speed digital encoded signals, supplying video television service.

Commercial LAN/WAN systems have been expanded to provide computer voice and video to their users. These operators did not compete with cable television systems for service. These systems were developed because the telephone system provided only telephone service, and cable television systems avoided commercial areas altogether.

Optical-fiber systems, developed mostly in telephone company laboratories, became an important contribution to the communications industry. Cable television distribution systems basically replaced the whole trunk cable plant with optical fiber. The telephone system replaced the long-line section of microwave and coaxial cable plant with optical fiber. Submarine optical cable provides telephone and digital services to Europe, Asia, and many other locations. Now, with optical-fiber systems practically at our door, HDTV and high-speed Internet can be realized as practical. Because the telephone system today consists of AT&T and the regional Bell operating companies as well as the spin-off companies, and they can offer nearly the same services as the cable system operators, which makes for a lot of competition. The interconnection of the main carrier of the Internet (telephone system) and the cable system was not difficult because optical-fiber systems were used by both networks. At present, the telephone system consists of optical-fiber plant feeding copper-based metallic wire plant to the customer. The cable plant consists of optical fiber for the bidirectional trunk plant and to the telephone plant, and uses coaxial cable plant to the subscribers. When either system extends optical-fiber plant to the home or office, the advantage of high-speed communications will be available to the consumer.

Questions

1. Explain the difference between telephone system local lines and trunk lines.

2. What is the purpose for twisting the pairs of wires in a telephone cable?

3. What is the purpose of placing loading coils (inductances) in telephone line circuits?

4. What does the acronym CATV mean?

5. Describe the network architecture for cable television systems.

6. Describe cable system choices of method for the reverse of upstream network.

7. Describe how optical-fiber technology is used in the forward or downstream sections of a cable television plant.

8. Explain an HFC network.

9. What is the largest network protocol for LAN networks?

10. Describe the three choices of network topology for LAN systems.

11. What type of network topology is used for Ethernet?

12. Explain the process used to transmit digital signals on a cable system.

APPENDIX A

Proof and Discussion of Pertinent Equations

1. The first two equations are explained using the simple resistance capacitance circuit where the input voltage source is in series with a resistor of R ohms in series with a capacitor of C microfarads. The output voltage is taken across the capacitor.

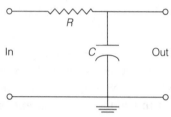

This circuit, in many applications, represents a voltage source where R is the equivalent resistance and C is the Miller capacitance of the amplifier input connected across the capacitance C. When V_{in} is a sinusoidal signal voltage, the a.c. value of the capacitive reactance in ohms is

$$X_c = \frac{1}{2\pi f C} \quad \text{when the magnitude of } X_c = R \text{ in solving for } f$$

Substituting $R = 1/2\pi f C$,

$$f = \frac{1}{2\pi RC}$$

where f is the frequency of the input voltage V_{in} (RMS) when $X_c = R$. This is called the cut-off frequency of this simple low-pass filter, which is often referred to as a basic integrator. The attenuation in output voltage at this frequency may be calculated as follows using the voltage divider theorem:

$$V_{out} = \frac{Xc}{RjX_c}(V_{in}) \quad \text{where } RjX_c \text{ is } 90° \text{ leading}$$

Converting to polar coordinates and solving,

$$\frac{V_{out}}{V_{in}} = \frac{X_c}{(R^2 + X_c^2)\arctan\dfrac{X_c}{R}}$$

Since $X_c = R$,

$$\frac{V_{out}}{V_{in}} = \frac{|X_c|}{(2X_c^2)^{1/2}} = \frac{|X_c|}{(X_c)(2)^{1/2}} = \frac{1}{(2)^{1/2}} = 0.707$$

$$V_{out} = 0.707V_{in} \quad \text{or} \quad \frac{V_{out}}{V_{in}} = 0.707$$

Since V_{out}/V_{in} represents the circuit gain,

$$20 \log \frac{V_{out}}{V_{in}} = 20 \log 0.707 = -3 \text{ dB}$$

where the minus sign means a negative gain or attenuation.

2. Proof of the rise time equation for the same resistor capacitive circuit. Since digital signal voltages vary between two values, the response of this circuit to sudden voltage changes is important. The rise time is the measure of this response. A short rise time means this circuit responds faster to sudden changes in input voltage. This circuit follows the charging curve for the RC circuit where $R \times C$ = the circuit time constant where R is in ohms and C is in farads.

Since the output voltage is V_c, we may write

$$V_c = V_{max}(1 - e^{-t/RC}) \quad \text{where } RC = \text{time constant}$$

The IEEE stated that the rise time is the time it takes for V_c at $0.1V_{max}$ to reach V at $0.9V_{max}$.

$$V_c = 0.1V_{max}$$

and substituting,

$$0.1V_{max} = V_{max}(1 - e^{-t/RC})$$

Multiplying,

$$0.1V_{max} = V_{max} - V_{max}e^{-t/RC}$$

Combining terms,

$$-0.9V_{max} = -V_{max}e^{-t/RC}$$

Dividing by V_{max} and multiplying by -1,

$$-0.9 = e^{-t/RC}$$

Taking the natural logarithm of both terms,

$$\ln - 0.9 = \ln e^{-t/RC}$$

Using the calculator,

$$-0.9 = \frac{-t}{RC}$$

So $t = 0.1RC$ at $0.1V_{max}$ when $V_c = 0.9V_{max}$. Substituting,

$$0.9V_{max} = V_{max}(1 - e^{-t/RC})$$

Multiplying,

$$0.9V_{max} = V_{max} - V_{max}e^{-t/RC} - 0.1V_{max}$$

$$= -V_{max}e^{-t/RC}$$

Dividing by V_{max}, multiplying by -1, and rearranging terms,

$$e^{-t/RC} = 0.1$$

Taking the natural logarithm,

$$\ln e^{-t/RC} = \ln 0.1$$

Solving for t,

$$\frac{t}{RC} = 2.3 \text{ at } 0.9V_{max}$$

The rise time is defined as the time it takes for the capacitor to charge from 10% V_{max} to 90% V_{max}.

$$\text{Rise time } t_r = t_{0.9} - t_{0.1}$$

$$= 2.3RC - 0.1RC = 2.2RC$$

$$t_r = 2.2RC$$

To investigate the frequency response of this basic low-pass filter, recall that we previously derived

$$f_c = \frac{1}{2\pi RC}$$

Solving for the RC product,

$$RC = \frac{1}{2\pi f_c}$$

Substituting for $RC = t_r/2.2$,

$$RC = \frac{t_r}{2.2} = \frac{1}{2\pi f_c}$$

$$\frac{2.2}{2\pi f_c} = \frac{0.35}{f_c}$$

Solving for f_c,

$$f_c = \frac{0.35}{t_r}$$

Since the bandwidth (BW) is from zero frequency to the cut-off frequency,

$$f_c = \text{BW}$$

Substituting,

$$\text{BW} = \frac{0.35}{t_r}$$

This result relates the low-pass filter circuit bandwidth to the rise time. A shorter (quicker rise time) increases the bandwidth.

3. Proof of the equation

$$\frac{P_r}{P_t} = G_t + G_r - 32.5 \text{ dB} - 20 \log d_{\text{km}} - 20 \log f_{\text{MHz}}$$

Consider a source of power at a point in space radiating electromagnetic energy into space in all directions. the equation describing this action is

$$P_d = \frac{P_t}{4\pi d^2}$$

where

P_d = the power density in watts per square meter (w/m²) at distance d meters from source P_t

P_t = the transmitted power in watts

d = the distance from the point source in space

$4\pi d^2$ = the area of a sphere in square meters

The receive power at a receiving antenna is given as

$$P_r = A_{\text{eff}} P_d$$

where A_{eff} is the effective area of the receive antenna in square meters.

$$A_{\text{eff}} = \frac{\lambda^2 G_r}{4\pi}$$

where

λ = the wavelength at operating frequency

G_r = the gain of receive antenna over an isotropic source

Note: an isotropic source is the point source in space and physically does not exist.

The simplest actual antenna is a dipole antenna that exhibits a gain of 1.64 over the theoretical point source. Substituting,

$$P_r = \frac{(\lambda^2 G_r)(G_t P_t)}{(4\pi)(4\pi d^2)}$$

where G_t is the gain of the transmitting antenna over the point source (isotropic). Solving for the ratio of the receive power to the transmitting power, we may write

$$\frac{P_r}{P_t} = \frac{\lambda^2 G_r G_t}{16\pi^2 d^2}$$

Recall $\lambda = c/f$ where $c = 3 \times 10^8$ m/s (the speed of electromagnetic wave propagation). Solving for f,

$$f = \frac{c}{\lambda}$$

and substituting,

$$\frac{P_r}{P_t} = \frac{c^2 G_r G_t}{16\pi^2 d^2 f^2}$$

Rewriting the preceding power ratio expression in dB form,

$$\frac{P_r}{P_t} = (20 \log c + G_r + G_t) - (20 \log 4 + 20 \log \pi + 20 \log d + 20 \log f)$$

Because the distance between the transmission point and the receive point is often large, it is practical to express d in kilometers (km). Also, because the operating frequency is high it is also practical to express f in megahertz (MHz). Converting the previous equation for d (km) and f (MHz) and substituting the constant values and combining, we may write

$$\frac{P_r}{P_t} = (20 \log 3 + 20 \log 10^8 + G_r + G_t) - (20 \log 4 + 20 \log \pi + 20 \log d + 20 \log 10^3 + 20 \log f + 20 \log 10^6)$$

$$= (9.5 \text{ dB} + 160 \text{ dB} + G_r + G_t) - (12 \text{ dB} + 10 \text{ dB} + 20 \log d + 60 \text{ dB} + 20 \log f + 120 \text{ dB})$$

$$= (169.5 \text{ dB} + G_r + G_t) - (202 \text{ dB} + 20 \log d + 20 \log f)$$

$$= G_r + G_t - 32.5 \text{ dB} - 20 \log d - 20 \log f$$

This equation is simple to use because the receive and transmitter antenna gains are published in dBi and so can be directly substituted in the preceding equation. This equation can be used for satellite, line-of-site microwave, and any RF point-to-point applications. The negative terms in the preceding equation represent the path loss between the transmit and receive locations.

4. Proof of the equation

$$E_r = 0.021 V_r f$$

where

E_r = the electric field intensity in microvolts per meter

V_r = the voltage induced in a dipole antenna

f = the operating frequency in megahertz

This equation is particularly useful in predicting the voltage level induced in a dipole test antenna in a field of E_r volts/meter.

Recall from derivation 3, the power density (in W/m²) can be calculated as

$$P_d = \frac{P_t}{4\pi d^2}$$

It can be shown that

$$P_d = \frac{E_r^2}{377}$$

where E_r is in v/m:

$$377 = \text{impedance of free space, in ohms}$$

Also recall the effective area of an antenna is

$$A_{\text{eff}} = \frac{G\lambda^2}{4\pi} \quad \text{in square meters}$$

where

G = the antenna gain over an isotropic (point source)

λ = the wavelength at the operating frequency

The power density at distance d in meters should produce a power into a receive antenna.

$$P_{\text{ant}} = \frac{V_{\text{rec}}^2}{Z_o}$$

where Z_o is the impedance of the receive antenna.

Combining equations,

$$\frac{V_{\text{rec}}^2}{Z_o} = \frac{E_r^2}{377}$$

$$A_{\text{eff}} = \frac{\lambda^2 G_r}{4\pi}$$

so

$$\frac{V_{rec}^2}{Z_o} = \frac{E_r^2 G \lambda^2}{377(4\pi)}$$

Because $377 = 120\pi$, we can substitute

$$\frac{V_{rec}^2}{Z_o} = \frac{E_r^2 G \lambda 2}{480\pi^2}$$

Solving for E_r^2,

$$E_r^2 = \frac{480\pi^2 V_{rec}^2}{Z_o G \lambda^2}$$

$$m = \frac{300 \text{ m/s}}{F(\text{in MHz})}$$

Substituting,

$$E_r = \left(\frac{480}{1.64(72)}\right)^{1/2} \frac{\pi V_{rec} f_{MHz}}{300}$$

Taking the square root,

$$E_r = \left(\frac{480}{GZ_o}\right)^{1/2} \frac{\pi V_{rec} f_{MHz}}{300} \text{ volts}$$

The test antenna is a half-wave dipole with a gain of 1.64. The impedance of the half-wave dipole antenna is 72 ohms.

$$E_r = \left(\frac{480}{1.64(72)}\right)^{1/2} \frac{\pi V_{rec} f_{MHz}}{300}$$

Calculating,

$$E_r = (2)(0.01) f_{MHz} V_{rec}$$

$$= 0.02 f_{MHz} V_{rec}$$

where E_r is in microvolts per meter and V_{rec} is the antenna voltage in microvolts per meter.

This equation is used to test a signal level in microvolts using a half-wave dipole test antenna at a distance from the source in a field intensity of E_r in microvolts/meter.

APPENDIX B

Transmission Line Calculations

Velocity of Propagation

$$\text{Velocity} = \frac{\text{distance}}{\text{time}}$$

Velocity of propagation of electrical energy along a transmission line of length D meters is given by

$$v_p = \frac{D}{\sqrt{LC}}$$

L = inductance of line, in henrys

C = capacitance of line, in farads

v_p = velocity of propagation, in meters per second

Characteristic Impedance of a Transmission Line

A schematic diagram of a section transmission line follows.

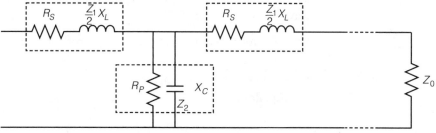

R_S is a.c. series resistance caused by wires, in ohms.

X_L is a series inductive reactance caused by twisted wires.

R_P is insulation leakage resistance between conductors.

X_C is capacitive reactance caused by capacitance between conductors.

It can be shown that

$$Z_O = \sqrt{Z_1 Z_2}$$

where Z is the impedance of two sets of R_S in series with X_L. Also, it can be shown that if $R_S = 0$ $R_P = \infty$ for the lossless transmission line

$$Z_O = \sqrt{\frac{L}{C}}$$

Therefore, as series inductance is added by loading the pair, the characteristic impedance of the line is increased. Given the following transmission line,

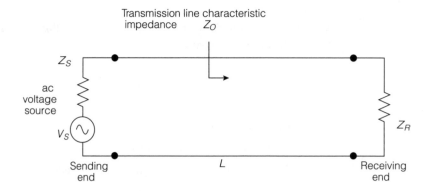

if Z_S and Z_R do not equal Z_O, voltage and current reflections will result along the transmission line length L. The reflection coefficient K is defined as the magnitude of the reflected voltage wave to the incident voltage wave.

V^- RMS voltage of the reflected wave, reflected from the mismatch impedance

V^+ RMS voltage of the incident wave going toward the mismatch impedance

$$K = \frac{|V^-|}{|V^+|}$$

Example: If $V^- = 0$ V, there is no reflected wave and $K = 0$. K in terms of the impedance mismatch is represented as

$$\text{at the sending end } K_S = \frac{Z_S - Z_O}{Z_S + Z_O}$$

$$\text{at the receiving end } K_R = \frac{Z_R - Z_O}{Z_R + Z_O}$$

Example: For a perfect impedance match,

$$Z_O = Z_R = Z_S$$

$$K_S = 0 \, K_R, \text{ or } 0$$

meaning no reflections.

If V^+ is the incident voltage wave traveling down the line and V^- is the reflected wave traveling up the line caused by the mismatch of the receiving end impedance, V_{MAX} occurs when they reinforce each other to cause a voltage maximum.

$$|V_{MAX}| = |V^+| + |V^-| \qquad \textbf{(B-1)}$$

$$|V_{MIN}| = |V^+| - |V^-| \qquad \textbf{(B-2)}$$

V^+ and V^- cancel each other out. The voltage standing-wave ratio is defined as

$$\text{VSWR} = \frac{|V_{MAX}|}{|V_{MIN}|}$$

Substituting in Equations B-1 and B-2,

$$\text{VSWR} = \frac{|V^+| + |V^-|}{|V^+| - |V^-|} \qquad \textbf{(B-3)}$$

and multiplying Equation B-3 by $1/|V^-|$,

$$\text{VSWR} = \frac{1 + \dfrac{|V^-|}{|V^+|}}{1 - \dfrac{|V^-|}{|V^+|}} \quad \text{and} \quad K = \frac{|V^-|}{|V^+|}$$

so

$$\text{VSWR} = \frac{1 + K}{1 - K}$$

K can be K_R or K_S. Return loss is a measure of the incident wave power to the reflected wave power. Mathematically,

$$\text{RL} = 10 \log \frac{|P+|}{|P-|}$$

and

$$|P^+| = \frac{|V^+|^2}{Z_O} \quad \text{and} \quad |P^-| = \frac{|V^-|^2}{Z_O}$$

Substituting,

$$RL = 10 \log \frac{\dfrac{|V^+|^2}{Z_O}}{\dfrac{|V^-|^2}{Z_O}} = 10 \log \frac{|V^+|^2}{|V^-|^2}$$

$$RL = 20 \log \frac{|V^+|}{|V^-|}$$

Since $K = |V^-|/|V^+|$,

$$RL = 20 \log \frac{1}{K}$$

Return loss is the difference expressed in dB between the incident and reflected waves along a transmission line.

APPENDIX C

Coaxial Cable System Powering

A section of cable plant is shown; the resistances of cable sections and the distances between amplifiers are noted.

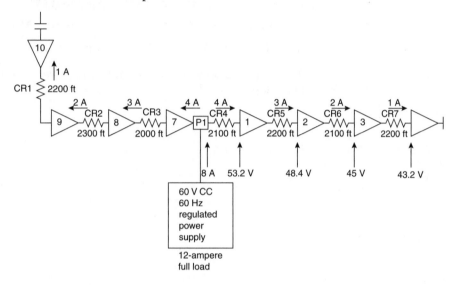

—|⊢— Power block signal pass

Cable section	Length (in 1000 ft)	Resistance (in ohms)*	Current (in amps)	Voltage drop (in volts)†
CR1	2.2	1.8	1	1.8
CR2	2.3	1.8	2	3.6
CR3	2	1.6	3	4.8
CR4	2.1	1.7	4	6.8
CR5	2.2	1.8	3	4.8
CR6	2.1	1.7	2	3.4
CR7	2.2	1.8	1	1.8

*0.8 ohms per 1000 ft

†Resistance × current

Cable loop resistance is 0.8 ohm per thousand feet. The amplifiers draw 1 ampere each and the minimum allowable amplifier voltage is 40 volts. The power supply delivers 8 amps.

Now the voltage at each amplifier location can be calculated and labeled on the diagram.

Amp 1: 60 V − 6.8 V = 53.2 V   Amp 7: No voltage drop

Amp 2: 53.2 V − 4.8 V = 48.4 V   Amp 8: 60 V − 4.8 = 55.2 V

Amp 3: 48.4 V − 3.4 V = 45 V   Amp 9: 55.2 V − 3.6 V = 51.6 V

Amp 4: 45 V − 1.8 V = 43.2 V   Amp 10: 51.6 V − 1.8 V = 49 V

All amplifiers have proper voltage, and power supply is only 75 percent fully loaded.

APPENDIX D

Broadband Noise Combining

This appendix discusses the noise-combining effect in single cable reverse cable television systems. In a subsplit reverse system, carrier C_1 is at channel T-7 (7.0 MHz) and C_2 is at channel T-8 (13 MHz). These channels are combined in a splitter and a directional coupler with equal noise levels.

Channel T-7 $C_1 = +20$ dBmV carrier level $N_1 = -25$ dBmV noise level
Channel T-8 $C_2 = +20$ dBmV carrier level $N_2 = -25$ dBmV noise level

$C_1 = +20$ dBmV $C_1/N_1 = 45$ dB   At output
$N_1 = -25$ dBmV

                                $C_1 = +17$ dBmV
                                $N_1 = -28$ dBmV $C_1/N_O = 42$ dB

$C_2 = +20$ dBmV
$N_2 = -25$ dBmV   Combiner / splitter
                                 $C_2 = +20$ dBmV
                                 $N_2 = -25$ dBmV

                $C_2/N_2 = 45$ dB $C_2/N_O = 42$ dB

Since noise level is constant across the frequency band, the output noise $N_O = -28$ dBmV $+ 3$ dBmV $= -25$ dBmV.

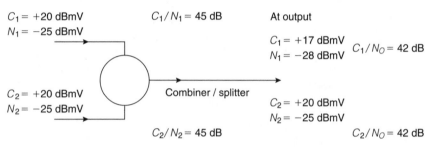

                              At output
                                 $C_1/N_O = +19$ dBmV $- (-25.4$ dB$) = 44.5$ dB

               $C_1/N_1 = 45$ dB $C_1 = +19$ dBmV
$C_1 = +20$ dBmV $N_1 = -26$ dBmV
$N_1 = -25$ dBmV -1 dB

            -10 dB $C_2/N_O = +10$ dBmV $- (-25.4$ dB$) = 35.5$ dB
$C_2 = +20$ dBmV   Directional coupler
$N_2 = -25$ dBmV $C_2 = +10$ dBmV
                               $N_2 = -35$ dBmV
                $C_2/N_2 = 45$ dB

To combine N_1 and N_2 at the output on a power basis,

$$-26 = 10 \log N_1 \quad \log N_1 = -2.6 \quad N_1 = 0.00256$$

$$26 = 10 \log N_2 \quad \log N_2 = -3.5 \quad N_2 = 0.00032$$

$$\log N_O = N_O \quad N_O = 0.00288$$

$$N_O = 10 \log N_O = 10 \log 0.00288 = -25.4 \text{ dB}$$

Summary: In the case of the two-port combiner, carrier-to-noise level for both C_1 and C_2 was decreased by 3 dB. For the case of the -10-dB directional combiner where C_2 is combined with C_1 through the -10-dB port, C_1 only had a ½-dB decrease in carrier-to-noise level, while C_2 had a 9½-dB decrease in carrier-to-noise level.

APPENDIX E

Cascaded Amplifier Theory

Noise voltage generated by a 75-ohm resistor at 20° C, expressed in dBmV. This voltage appears across the input resistance of the first 75-ohm amplifier in a cascade of repeater amplifiers.

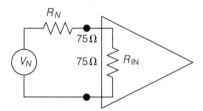

$$V_N = \sqrt{4KTBR}$$

K = Boltzmann's constant, $1.38 \times 10^{-23} \, \text{J/K}$

T = temperature, in K

B = frequency bandwidth, in Hz

R_N = resistance, in ohms

At 20° C, $T = 20°\,\text{C} + 273°\,\text{C} = 293\,\text{K}$. For a video bandwidth of 4 MHz, $B = 4 \times 10^6$ Hz

$$V_N = \sqrt{4 \times 293 \times 1.38 \times 10^{-23} \times 4 \times 10^6 \times 75}$$

$$= \sqrt{485{,}208 \times 10^{-17}} = \sqrt{4.85 \times 10^{-12}} = 2.2 \times 10^{-6} \, \text{V}$$

$$= 2 \, \mu\text{V}$$

This is the noise V_N generated by R_N; it appears across a circuit of two 75-ohm resistors connected in series.

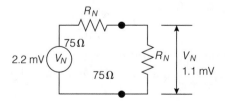

V_{IN} is that portion of noise voltage that appears across the amplifier input terminal and is $\frac{1}{2}V_N$ according to the voltage divider principle. In dBmV,

$$20 \log \frac{1.1 \times 10^{-6} \text{ V}}{1 \times 10^{-3} \text{ V}} = 20 \log 1.1 \times 10^{-3}$$

$$= 20 \times -2.959 = -59.2 = -59 \text{ dBmV}$$

This level in dBmV constitutes the so-called noise floor in cable television amplifier cascades.

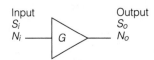

S_i = signal input power level

N_i = noise input power level

S_o = signal output power level

N_o = noise output power level

G = gain of amplifier

For the amplifier, the noise factor is given as

$$F = \frac{\dfrac{S_i}{N_i}}{\dfrac{S_o}{N_o}} = \frac{\text{Input signal-to-noise ratio}}{\text{Output signal-to-noise ratio}}$$

Rewriting,

$$F = \frac{S_i}{N_i} \times \frac{N_o}{S_o}\left(\frac{N_o}{N_i}\right)\left(\frac{S_i}{S_o}\right)$$

Since $G = S_o/S_i$

$$\frac{S_i}{S_o} = \frac{1}{G}$$

Substituting,

$$F = \frac{N_o}{N_i}\left(\frac{1}{G}\right) \tag{E-1}$$

The noise power output of an amplifier consists of the input noise level amplified G times, plus the noise N_n generated by the amplifier itself. Mathematically,

$$N_o = N_i(G) + N_n \tag{E-2}$$

Combining Equations E-1 and E-2,

$$F = \frac{N_i(G) + N_n}{N_i(G)} = 1 + \frac{N_n}{N_i(G)} \tag{E-3}$$

Consider a cascade of three amplifiers 1, 2, and 3.

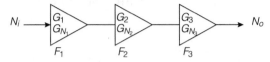

For the cascade,

$$F_c = \frac{N_o}{N_i(G_c)}$$

where G_c is gain of cascade. No noise or power output can be calculated.

$$N_o = \underset{\substack{\text{Cascade of}\\\text{input noise}}}{N_i(G_1)(G_2)(G_3)} + \underset{\substack{\text{Noise}\\\text{generated}\\\text{by amplifier 1}}}{N_{n_1}(G_1)(G_2)} + \underset{\substack{\text{Noise}\\\text{generated}\\\text{by amplifier 2}}}{N_{n_2}(G_3)} + \underset{\substack{\text{Noise}\\\text{generated}\\\text{by amplifier 3}}}{N_{n_3}} \tag{E-4}$$

Also,

$$G_c = (G_1)(G_2)(G_3) \tag{E-5}$$

For the cascade, substituting Equation E-3 in Equation E-1,

$$F = \frac{N_i(G_1)(G_2)(G_3) + N_{n_1}(G_2)(G_3) + N_{n_3}}{N_i(G_1)(G_2)(G_3)} \tag{E-6}$$

Rewriting Equation E-6,

$$F = 1 + \frac{N_{n_1}}{N_i G_1} + \frac{N_{n_2}}{N_i(G_1)(G_3)} + \frac{N_{n_3}}{N_i(G_1)(G_2)(G_3)}$$

and

$$F_2 = 1 + \frac{N_{n_2}}{N_i G_2} \qquad F_3 = 1 + \frac{N_{n_3}}{N_i G_3}$$

Rearranging the above for F_1, F_2, and F_3,

$$F_1 - 1 = \frac{N_{n_1}}{N_i G_1} \qquad F_2 - 1 = \frac{N_{n_2}}{N_i G^2} \qquad F_3 - 1 = \frac{N_{n_3}}{N_i G_3}$$

Substituting back into Equation E-6, for the cascade

$$F = F_1 + \frac{F_2 - 1}{G_1} + \frac{F_3 - 1}{G_1 G_2}$$ (E-7)

$$F = 10 \log F$$ (E-8)

where F is the noise figure, in dB. Therefore, Equation E-7 for N stages becomes:

$$F = F_1 + \frac{F_2 - 1}{G_1} + \frac{F_3 - 1}{G_1 G_2} + \cdots + \frac{Fn - 1}{G_1 G_2 \ldots G_{n-1}}$$

Given: a two-amplifier cascade connected by a piece of cable with its loss equal to an amplifier's gain. The cable will generate noise and has negative gain.

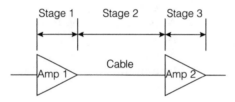

$F_1 = 10$ dB $F_2 = 10$ dB $F_3 = 10$ dB

$G_1 = 10$ dB $G_2 = -10$ dB $G_3 = 10$ dB

Typically, F and G are given in dB form, so in order to use Equation E-7, F and G have to be converted to power ratios. For amplifier 1,

$$10 = 10 \log F_1 \quad \log F_1 = 1 \quad F_1 = \log^{-1} \quad F_1 = 10$$

In like manner, $G_1 = 10$. For the cable,

$$F_2 = 10$$

$$-10 = 10 \log G_2 \quad \log G_2 = -1 \quad G_2 = \log^{-1}(-1) \quad G_2 = 0.1$$

For amplifier 2,

$$F_2 = 10 \quad \text{and} \quad G_2 = 10 \text{ (same as amplifier 1)}$$

Substituting,

$$F = F_1 + \frac{F_2 - 1}{G_1} + \frac{F_3 - 1}{G_1 G_2}$$

$$F_c = 10 + \frac{10 - 1}{10} + \frac{10 - 1}{10(0.1)} = 10 + 0.9 + 9 = 19.9 = 20$$

$$F_{c_{dB}} = 10 \log F_c = 10 \log 20 = 10(1.3) = 13 \text{ dB}$$

Rule: Doubling the number of amplifiers increases noise by 3 dB. This rule can be applied to a case of many amplifiers in cascade—32, in this example. Because the output signal level is constant and the buildup of noise is increased by 3 dB every time the number of amplifiers is doubled, the signal-to-noise ratio expressed in dB decreases by 3 dB every time the number of amplifiers is doubled.

Since the signals carried on an amplifier cascade are television carriers, for a C/N at amplifier 1 with an output of 60 dB, consider the following example.

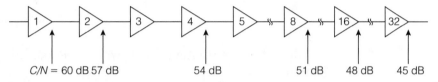

$C/N = 60$ dB 57 dB   54 dB   51 dB   48 dB   45 dB

The carrier-to-noise ratio is degraded 3 dB each time the amplifier number in the cascade is doubled. A C/N of 45 dB for present-day cable systems is too low. A C/N of 49 dB is acceptable. A test can be performed to find the C/N of a single amplifier. But first consider that for television carrier signal C_i input level and C_o output level,

$$F = \frac{C_i}{N_i} \times \frac{N_o}{C_o}$$

Rewriting in terms of the output carrier-to-noise level C_o/N_o,

$$\frac{C_o}{N_o} \times F = \frac{C_i}{N_i} \times \frac{N_o}{C_o} \times \frac{C_o}{N_o}$$

$$\frac{C_o F}{N_o} = \frac{C_i}{N_i}$$

Dividing both sides by F,

$$\frac{C_o}{N_o} = \frac{C_i}{N_i F} \qquad \text{(E-9)}$$

Rewriting Equation E-7 in terms of dB,

$$\frac{C_o}{N_o} \text{ (in dB)} = C_{i_{dB}} - (N_{i_{dB}} + F_{dB})$$

$$= \text{signal input level (dBmV)} - (-59 \text{ dBmV})$$

$$\frac{C}{N} = 59 - F_{\text{dB}} + \text{signal input dBmV dB for a single amplifier}$$

The test configuration for the noise figure of an amplifier under test is shown.

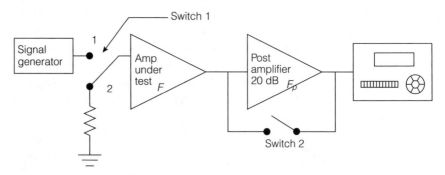

With switch 1 in position 1 and switch 2 closed, gain of amplifier is measured as

$$\text{Gain} = \frac{C_o}{C_i}$$

Signal level meter with noise test feature will not need any correction factor. Or, in terms of dB,

$$C_o - C_i = \text{gain} = 22 \text{ dB}$$

Switch 1 is placed in position 2 to terminate amplifier input. Switch 2 is opened to allow the output noise of amplifier to be further amplified 22 dB by the post amplifier. Output noise is read on signal level meter as -20 dBmV.

Actual output noise is calculated by

$$\text{Measured output noise} = \text{Post amp gain} + F + \\ \text{Input noise} + F_p + \text{Amplifier gain} - 20 \text{ dBmV}$$

$$= 20 \text{ dB} + F - 59 \text{ dBmV} + 22 \text{ dB} + 7 \text{ dB} - 20 \text{ dBmV}$$

$$= -59 \text{ dBmV} + 49 \text{ dB} + F$$

$$F = 10 \text{ dB noise contributed}$$

Because the noise figure F has been found for a single amplifier, the C/N for a single amplifier can also be calculated. Example: For an amplifier with a normal carrier input level of $+10$ dBmV and a noise figure F of 10 dB,

$$\frac{C_o}{N_o} = 59 - F + C_i$$

$$= 59 - 10 + 10$$

$$= 59 \text{ dB}$$

Typically, CATV amplifiers have noise figures better than 10 dB, usually 4 to 7 dB. For the case of identical amplifiers connected together by lengths of cable with a loss equal to each amplifier gain, it was seen that the noise increased by 3 dB every time the number of amplifiers in cascade was doubled.

$$F_o = F_1 + 10 \log n$$

Where F_1 is the noise figure of a single amplifier and n is the number of amplifiers in cascade. For two amplifiers,

$$10 \log 2 = 10(0.301) = 3 \text{ dB}$$

For four amplifiers,

$$10 \log 4 = 10(0.602) = 6 \text{ dB}$$

For eight amplifiers,

$$10 \log 8 = 10(0.903) = 9 \text{ dB}$$

Using this mathematical formula, the noise figure for a cascade of seven amplifiers can be calculated. If F_1 for a single amplifier is 5 dB, then

$$F_o = 5 \text{ dB} + 10 \log 7 = 5 \text{ dB} + 8.5 \text{ dB} = 13.5 \text{ dB}$$

In the example of the cascade of 32 amplifiers, the C/N decreased by 3 dB every time the number of amplifiers in the cascade doubled. So we can write

$$\frac{C}{N} = \frac{C}{N} - 10 \log n$$

$10 \log n$ is called the cascade factor. At the first amplifier, $C/N = 60$ dB. Therefore, at the eighth amplifier,

$$\frac{C}{N} = 60 - 10 \log 8 = 60 - 9 = 51 \text{ dB}$$

The C/N calculation is important because the signal quality of the amplifier cascade depends on a high signal-to-noise ratio. If the first amplifier becomes noisy, the whole cascade suffers. Example: For the 32nd amplifier, the C/N has degraded to $50 - 15 = 35$ dB. This C/N is totally unacceptable. A temporary solution would be to switch the first and last amplifier. Now only service on the end will be affected and not much at that.

Third-Order Distortion

In a similar proof, the buildup of amplifier distortion can be shown. Amplifier distortion appears as second- and third-order distortion. Second-order distortion involves two frequency components such as the second harmonics of a carrier. Third-order distortion involves three frequency components.

- Intermodulation $2f_1 \pm f_2$ $\quad 2f_2 \pm f_1$
- Cross-modulation is caused by any false carrier generated by any third-order distortion with another carrier's modulating signal, which affects any of the correct signal carriers. This is similar to another channel's signal affecting the viewed channel.
- Third harmonic, represented by $3f_1$, $3f_2$, etc.

Distortion Buildup for a Cascade

For second-order distortion, the cascade calculation is like that for noise.

Second-order distortion for the cascade =
second-order for a single amplifier $-10 \log n$

For third-order distortion, i.e., cross-modulation, single triple beat (third harmonic), or composite triple beat,

Third-order distortion for the cascade =
third-order for a single amplifier $-20 \log n$

Manufacturers of cable amplifiers specify call of the single amplifier parameters. There the calculations using the formulas can be used to predict the amplifier cascade performance.

Historically, the use of push-pull-type circuitry all but eliminated second-order distortion. Thus, third-order distortion became the limiting factor in setting the maximum number of amplifiers in cascade. Noise buildup is still an important specification.

APPENDIX F

Answers to Odd-Numbered Problems

Chapter 1

1. -40 dB

3. -26 dB

5. Gain needed: 20 dB

Chapter 2

1. $+45$ dBmV

3. $+37.5$ dBmV

5. 54 dB

7. 46 dB (not acceptable)

9. 24.35 in, 24 5/8 in

Chapter 3

1. 5.62 ft, or approximately 6 feet of separation

3. Channel 1: 37 dBmV; channel 2: $+36$ dBmV; channel 3: $+35$ dBmV; channel 8: $+30$ dBmV. These are the required input levels to produce a $+22$ dBmV output level.

5. In an eight-input combining network, using ideal signal splitters, $+31$ dBmV at all eight inputs will combine to a constant $+22$ dBmV output.

Chapter 4

1. The optical fiber is approximately 100 times better (that much lower loss) compared with the coaxial cable of the same length.

3. -11 dBm

5. -3.15 dBm transmit level

Chapter 5

1. 0101 is the digital number

3. 0.648 microsecond

Chapter 6

1. +7 dBmV

3. A staple nail driven through the cable.

Chapter 7

1. +3 dBmV for the audio carrier level

3. (a) +28 dBmV (b) −12 dBmV

5. First look for a blown fuse inside the distribution amplifier. If replaced fuse blows again, look for a tap that has been burned by lightning.

Chapter 8

No problems.

GLOSSARY OF COMMONLY USED ACRONYMS

Term	Meaning
ACATS	Advisory Committee on Advance Television Service
ADC	analog-to-digital converter
ADSL	asymmetrical digital subscriber line
AMI	alternate mark inversion, digital signal coding
ASCII	American Standard Code for Information Interchange
ATM	asynchronous transfer mode
B8ZS	binary eight-zero suppression
BPSK	binary phase shift keying
CAP	competitive access provider
CCIR	International Radio Consultative Committee
CCIS	common channel interoffice signaling
CCITT	International Telephone and Telegraph Consultative Committee
CMTS	cable modem termination system
CO	central office
CPU	central processor unit
CRC	cyclical redundancy checking
CSMA/CD	carrier sense multiple access collision detection
DAC	digital-to-analog converter
DACS	digital access and cross-connect system
DCE	data communications equipment
DDD	direct distance dialing
DDN	digital data network
DTE	data terminal equipment
DLC	digital line carrier
DOCSIS	data over cable service interface specification
DOS	disk operating system
DOV	data over voice

DS-0	digital signal, level 0
DS-1	digital signal, level 1
DSS	direct satellite service
DSP	digital signal processing
DSX-1	digital system cross connect
DVD	digital video disc
DVOD	digital video on demand
8-VSB	eight-level vestigial sideband
FCC	Federal Communications Commission
FDDI	fiber-distributed data interface
FDM	frequency division multiplex
FDMA	frequency division multiple access
FSK	frequency shift keying
FTTP	fiber to the premises
GPS	global positioning system
HDLC	high-level data link control
HDTV	high-definition television
HFC	hybrid fiber coax
IEEE	Institute of Electrical Electronics Engineers
IP	Internet provider
ISDN	integrated services digital network
ITU	International Telecommunications Union
JPEG	Joint Photographic Experts Group
LAN	local area network
LED	light-emitting diode
MPEG	Motion Picture Experts Group
MSO	multiple system operator
NOS	network operating system
OFDM	orthogonal frequency division multiplexing
OSI	open systems interconnection
PBX	private branch exchange
PCM	pulse code modulation
PCS	personal communications system

POD	point of deployment
PON	passive optical network
POP	point of presence
POTS	plain old telephone service
PSTN	public switch telephone network
QAM	quadrative amplitude modulation
QOS	quality of service
QPSK	quaternary phase–shift keying
RBOC	regional Bell operating company
RFID	radio frequency identification device
SCTE	Society of Cable Telecommunications Engineers
SDH	synchronous digital hierarchy
SDLC	synchronous data link control
SMPTE	Society of Motion Picture and Television Engineers
SONET	synchronous optical network
SPE	synchronous payload envelope
TDMA	time division multiple access
VPI	virtual path identifier
WAN	wide area network

REFERENCES

Bartlett, Eugene R. *Cable Communications*. New York: McGraw-Hill, Inc., 1995.

———. *Cable Television Handbook*. New York: McGraw-Hill, Inc., 2000.

———. *Cable Television Technology and Operations*. New York: McGraw-Hill, Inc., 1990.

Benson, K. Blair, ed. *Television Engineering Handbook*. New York: McGraw-Hill, Inc., 1996.

Chomycz, Bob. *Fiber Optic Installations*. New York: McGraw-Hill, Inc., 1996.

Freeman, Roger L. *Practical Data Communications*. New York: John Wiley and Sons, Inc., 1995.

———. *Telecommunication Transmission Handbook*. 2nd ed. New York: John Wiley and Sons, Inc., 1981.

Frenzel, Louis E. *Principles of Electronic Communication Systems*. 2nd ed. Glencoe, OH: McGraw-Hill, Inc., 2003.

Goralski, Walter. *Sonet/SDH*. New York: Osborne/McGraw-Hill, Inc., 2002.

Grant, William O. *Cable Television*. 2nd ed. Fairfax, VA: GWG Associates, 1998.

Hardy, William C. *VoIP Service Quality*. New York: McGraw-Hill, Inc., 2003.

Hecht, Jeff. *Understanding Fiber Optics*. 3rd ed. Upper Saddle River, NJ: Prentice Hall, 1998.

Hioki, Warren. *Telecommunications*. 3rd ed. Upper Saddle River, NJ: Prentice Hall, 1998.

Inglis, Andrew F. *Video Engineering*. New York: McGraw-Hill, Inc., 1993.

Noah, Jeffery. *FCC Required Measurements for U.S. Cable Systems*. Beaverton, OR: Tektronix, Inc.

Raskin, Donald, and Dean Stoneback. *Broadband Return Systems for Hybrid/Coax Cable TV Networks*. Upper Saddle River, NJ: Prentice Hall, 1997.

Robin, Michael, and Michel Poulin. *Digital Television Fundamentals*. New York: McGraw-Hill, Inc., 1998.

Sterling Jr., Donald J. *Technician's Guide to Fiber Optics*. 2nd ed. Albany, NY: Delmar Publishers, 1993.

Thomas, Jeffrey L. *Cable Television Proof-of-Performance.* Englewood Cliffs, NJ: Prentice Hall (IEEE Press), 1995.

Tunman, Ernest O. *Practical Multiservice LANS.* Canton, MA: Artech House, Inc., 1999.

Winch, Robert G. *Telecommunication Transmission Systems.* New York: McGraw-Hill, Inc., 1993.

INDEX